Wind Energy

STEM Road Map for Elementary School

Grade
5

Wind Energy

Grade
5

STEM Road Map
for Elementary School

Edited by Carla C. Johnson, Janet B. Walton, and
Erin Peters-Burton

National Science Teachers Association

Arlington, Virginia

National Science Teachers Association

Claire Reinburg, Director
Rachel Ledbetter, Managing Editor
Deborah Siegel, Associate Editor
Donna Yudkin, Book Acquisitions Manager

ART AND DESIGN
Will Thomas Jr., Director, cover and
interior design
Himabindu Bichali, Graphic Designer, interior
design

PRINTING AND PRODUCTION
Catherine Lorrain, Director

NATIONAL SCIENCE TEACHERS ASSOCIATION
David L. Evans, Executive Director
David Beacom, Publisher

1840 Wilson Blvd., Arlington, VA 22201
www.nsta.org/store
For customer service inquiries, please call 800-277-5300.

FSC
www.fsc.org
MIX
Paper from
responsible sources
FSC® C011935

NSTA is committed to publishing material that promotes the best in inquiry-based science education. However, conditions of actual use may vary, and the safety procedures and practices described in this book are intended to serve only as a guide. Additional precautionary measures may be required. NSTA and the authors do not warrant or represent that the procedures and practices in this book meet any safety code or standard of federal, state, or local regulations. NSTA and the authors disclaim any liability for personal injury or damage to property arising out of or relating to the use of this book, including any of the recommendations, instructions, or materials contained therein.

Library of Congress Cataloging-in-Publication Data
Names: Johnson, Carla C., 1969- editor. | Walton, Janet B., 1968- editor. | Peters-Burton, Erin E., editor. |
National Science Teachers Association, issuing body.
Title: Wind energy, grade 5 / edited by Carla C. Johnson, Janet B. Walton, and Erin Peters-Burton.
Description: Arlington, VA : National Science Teachers Association, [2017] | Series: STEM road map for elementary school | Includes bibliographical references and index.
Identifiers: LCCN 2017036769 (print) | LCCN 2017037984 (ebook) | ISBN 9781681404479 (e-book) |
ISBN 9781681404462 (print)
Subjects: LCSH: Wind power--Study and teaching (Elementary)
Classification: LCC TJ820 (ebook) | LCC TJ820 .W538 2017 (print) | DDC 372.35/8--dc23
LC record available at *https://lccn.loc.gov/2017036769*

The *Next Generation Science Standards* ("NGSS") were developed by twenty-six states, in collaboration with the National Research Council, the National Science Teachers Association and the American Association for the Advancement of Science in a process managed by Achieve, Inc. For more information go to *www.nextgenscience.org*.

CONTENTS

CONTENTS

ABOUT THE EDITORS AND AUTHORS

Dr. Carla C. Johnson is the associate dean for research, engagement, and global partnerships and a professor of science education at Purdue University's College of Education in West Lafayette, Indiana. Dr. Johnson serves as the director of research and evaluation for the Department of Defense–funded Army Educational Outreach Program (AEOP), a global portfolio of STEM education programs, competitions, and apprenticeships. She has been a leader in STEM education for the past decade, serving as the director of STEM Centers, editor of the *School Science and Mathematics* journal, and lead researcher for the evaluation of Tennessee's Race to the Top–funded STEM portfolio. Dr. Johnson has published over 100 articles, books, book chapters, and curriculum books focused on STEM education. She is a former science and social studies teacher and was the recipient of the 2013 Outstanding Science Teacher Educator of the Year award from the Association for Science Teacher Education (ASTE), the 2012 Award for Excellence in Integrating Science and Mathematics from the School Science and Mathematics Association (SSMA), the 2014 award for best paper on Implications of Research for Educational Practice from ASTE, and the 2006 Outstanding Early Career Scholar Award from SSMA. Her research focuses on STEM education policy implementation, effective science teaching, and integrated STEM approaches.

Dr. Janet B. Walton is the research assistant professor and the assistant director of evaluation for AEOP at Purdue University's College of Education. Formerly the STEM workforce program manager for Virginia's Region 2000 and founding director of the Future Focus Foundation, a nonprofit organization dedicated to enhancing the quality of STEM education in the region, she merges her economic development and education backgrounds to develop K–12 curricular materials that integrate real-life issues with sound cross-curricular content. Her research focuses on collaboration between schools and community stakeholders for STEM education and problem- and project-based learning pedagogies. With this research agenda, she works to forge productive relationships between K–12 schools and local business and community stakeholders to bring contextual STEM experiences into the classroom and provide students and educators with innovative resources and curricular materials.

Dr. Erin Peters-Burton is the Donna R. and David E. Sterling endowed professor in science education at George Mason University in Fairfax, Virginia. She uses her experiences from 15 years as an engineer and secondary science, engineering, and mathematics teacher to develop research projects that directly inform classroom practice in science and engineering. Her research agenda is based on the idea that all students should build self-awareness of how they learn science and engineering. She works to help students see themselves as "science-minded" and help teachers create classrooms that support student skills to develop scientific knowledge. To accomplish this, she pursues research projects that investigate ways that students and teachers can use self-regulated learning theory in science and engineering, as well as how inclusive STEM schools can help students succeed. During her tenure as a secondary teacher, she had a National Board Certification in Early Adolescent Science and was an Albert Einstein Distinguished Educator Fellow for NASA. As a researcher, Dr. Peters-Burton has published over 100 articles, books, book chapters, and curriculum books focused on STEM education and educational psychology. She received the Outstanding Science Teacher Educator of the Year award from ASTE in 2016 and a Teacher of Distinction Award and a Scholarly Achievement Award from George Mason University in 2012, and in 2010 she was named University Science Educator of the Year by the Virginia Association of Science Teachers.

Tamara J. Moore is an associate professor of engineering education in the College of Engineering at Purdue University. Dr. Moore's research focuses on defining STEM integration through the use of engineering as the connection and investigating its power for student learning.

Toni A. Sondergeld is an associate professor of assessment, research, and statistics in the School of Education at Drexel University in Philadelphia. Dr. Sondergeld's research concentrates on assessment and evaluation in education, with a focus on K–12 STEM.

Sandy Watkins is principal-in-residence for the Tennessee STEM Innovation Network. Ms. Watkins has been a science educator for kindergarten through postgraduate-level classes and has served as a STEM coordinator and consultant. She was also the founding principal of the Innovation Academy of Northeast Tennessee, which is a STEM school.

ACKNOWLEDGMENTS

This module was developed as a part of the STEM Road Map project (Carla C. Johnson, principal investigator). The Purdue University College of Education, General Motors, and other sources provided funding for this project.

See *www.routledge.com/products/9781138804234* for more information about *STEM Road Map: A framework for integrated STEM education*.

PART 1

THE STEM ROAD MAP

BACKGROUND, THEORY, AND PRACTICE

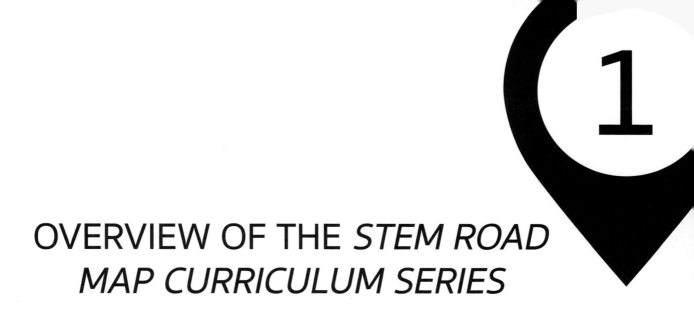

OVERVIEW OF THE *STEM ROAD MAP CURRICULUM SERIES*

Carla C. Johnson, Erin Peters-Burton, and Tamara J. Moore

The *STEM Road Map Curriculum Series* was conceptualized and developed by a team of STEM educators from across the United States in response to a growing need to infuse real-world learning contexts, delivered through authentic problem-solving pedagogy, into K–12 classrooms. The curriculum series is grounded in integrated STEM, which focuses on the integration of the STEM disciplines—science, technology, engineering, and mathematics—delivered across content areas, incorporating the Framework for 21st Century Learning along with grade-level-appropriate academic standards.

The curriculum series begins in kindergarten, with a five-week instructional sequence that introduces students to the STEM themes and gives them grade-level-appropriate topics and real-world challenges or problems to solve. The series uses project-based and problem-based learning, presenting students with the problem or challenge during the first lesson, and then teaching them science, social studies, English language arts, mathematics, and other content, as they apply what they learn to the challenge or problem at hand.

Authentic assessment and differentiation are embedded throughout the modules. Each *STEM Road Map Curriculum Series* module has a lead discipline, which may be science, social studies, English language arts, or mathematics. All disciplines are integrated into each module, along with ties to engineering. Another key component is the use of STEM Research Notebooks to allow students to track their own learning progress. The modules are designed with a scaffolded approach, with increasingly complex concepts and skills introduced as students progress through grade levels.

The developers of this work view the curriculum as a resource that is intended to be used either as a whole or in part to meet the needs of districts, schools, and teachers who are implementing an integrated STEM approach. A variety of implementation formats are possible, from using one stand-alone module at a given grade level to using all five modules to provide 25 weeks of instruction. Also, within each grade band (K–2, 3–5, 6–8, 9–12), the modules can be sequenced in various ways to suit specific needs.

STANDARDS-BASED APPROACH

The *STEM Road Map Curriculum Series* is anchored in the *Next Generation Science Standards* (*NGSS*), the *Common Core State Standards for Mathematics* (*CCSS Mathematics*), the *Common Core State Standards for English Language Arts* (*CCSS ELA*), and the Framework for 21st Century Learning. Each module includes a detailed curriculum map that incorporates the associated standards from the particular area correlated to lesson plans. The STEM Road Map has very clear and strong connections to these academic standards, and each of the grade-level topics was derived from the mapping of the standards to ensure alignment among topics, challenges or problems, and the required academic standards for students. Therefore, the curriculum series takes a standards-based approach and is designed to provide authentic contexts for application of required knowledge and skills.

THEMES IN THE *STEM ROAD MAP CURRICULUM SERIES*

The K–12 STEM Road Map is organized around five real-world STEM themes that were generated through an examination of the big ideas and challenges for society included in STEM standards and those that are persistent dilemmas for current and future generations:

- Cause and Effect

- Innovation and Progress

- The Represented World

- Sustainable Systems

- Optimizing the Human Experience

These themes are designed as springboards for launching students into an exploration of real-world learning situated within big ideas. Most important, the five STEM Road Map themes serve as a framework for scaffolding STEM learning across the K–12 continuum.

The themes are distributed across the STEM disciplines so that they represent the big ideas in science (Cause and Effect; Sustainable Systems), technology (Innovation and Progress; Optimizing the Human Experience), engineering (Innovation and Progress; Sustainable Systems; Optimizing the Human Experience), and mathematics (The Represented World), as well as concepts and challenges in social studies and 21st century skills that are also excellent contexts for learning in English language arts. The process of developing themes began with the clustering of the *NGSS* performance expectations and the National Academy of Engineering's grand challenges for engineering, which led to the development of the challenge in each module and connections of the module activities to the *CCSS Mathematics* and *CCSS ELA* standards. We performed these mapping processes with large teams of experts and found that these five themes provided breadth, depth, and coherence to frame a high-quality STEM learning experience from kindergarten through 12th grade.

Cause and Effect

The concept of cause and effect is a powerful and pervasive notion in the STEM fields. It is the foundation of understanding how and why things happen as they do. Humans spend considerable effort and resources trying to understand the causes and effects of natural and designed phenomena to gain better control over events and the environment and to be prepared to react appropriately. Equipped with the knowledge of a specific cause-and-effect relationship, we can lead better lives or contribute to the community by altering the cause, leading to a different effect. For example, if a person recognizes that irresponsible energy consumption leads to global climate change, that person can act to remedy his or her contribution to the situation. Although cause and effect is a core idea in the STEM fields, it can actually be difficult to determine. Students should be capable of understanding not only when evidence points to cause and effect but also when evidence points to relationships but not direct causality. The major goal of education is to foster students to be empowered, analytic thinkers, capable of thinking through complex processes to make important decisions. Understanding causality, as well as when it cannot be determined, will help students become better consumers, global citizens, and community members.

Innovation and Progress

One of the most important factors in determining whether humans will have a positive future is innovation. Innovation is the driving force behind progress, which helps create possibilities that did not exist before. Innovation and progress are creative entities, but in the STEM fields, they are anchored by evidence and logic, and they use established concepts to move the STEM fields forward. In creating something new, students must consider what is already known in the STEM fields and apply this knowledge appropriately. When we innovate, we create value that was not there previously and create new conditions and possibilities for even more innovations. Students should consider how their innovations might affect progress and use their STEM thinking to change current human burdens to benefits. For example, if we develop more efficient cars that use by-products from another manufacturing industry, such as food processing, then we have used waste productively and reduced the need for the waste to be hauled away, an indirect benefit of the innovation.

The Represented World

When we communicate about the world we live in, how the world works, and how we can meet the needs of humans, sometimes we can use the actual phenomena to explain a concept. Sometimes, however, the concept is too big, too slow, too small, too fast, or too complex for us to explain using the actual phenomena, and we must use a representation or a model to help communicate the important features. We need representations and models such as graphs, tables, mathematical expressions, and diagrams because it makes our thinking visible. For example, when examining geologic time, we cannot actually observe the passage of such large chunks of time, so we create a timeline or a model that uses a proportional scale to visually illustrate how much time has passed for different eras. Another example may be something too complex for students at a particular grade level, such as explaining the *p* subshell orbitals of electrons to fifth graders. Instead, we use the Bohr model, which more closely represents the orbiting of planets and is accessible to fifth graders.

When we create models, they are helpful because they point out the most important features of a phenomenon. We also create representations of the world with mathematical functions, which help us change parameters to suit the situation. Creating representations of a phenomenon engages students because they are able to identify the important features of that phenomenon and communicate them directly. But because models are estimates of a phenomenon, they leave out some of the details, so it is important for students to evaluate their usefulness as well as their shortcomings.

Sustainable Systems

From an engineering perspective, the term *system* refers to the use of "concepts of component need, component interaction, systems interaction, and feedback. The interaction of subcomponents to produce a functional system is a common lens used by all engineering disciplines for understanding, analysis, and design." (Koehler, Bloom, and Binns 2013, p. 8). Systems can be either open (e.g., an ecosystem) or closed (e.g., a car battery). Ideally, a system should be sustainable, able to maintain equilibrium without much energy from outside the structure. Looking at a garden, we see flowers blooming, weeds sprouting, insects buzzing, and various forms of life living within its boundaries. This is an example of an ecosystem, a collection of living organisms that survive together, functioning as a system. The interaction of the organisms within the system and the influences of the environment (e.g., water, sunlight) can maintain the system for a period of time, thus demonstrating its ability to endure. Sustainability is a desirable feature of a system because it allows for existence of the entity in the long term.

In the STEM Road Map project, we identified different standards that we consider to be oriented toward systems that students should know and understand in the K–12 setting. These include ecosystems, the rock cycle, Earth processes (such as erosion, tectonics, ocean currents, weather phenomena), Earth-Sun-Moon cycles, heat transfer, and the interaction among the geosphere, biosphere, hydrosphere, and atmosphere. Students and teachers should understand that we live in a world of systems that are not independent of each other, but rather are intrinsically linked such that a disruption in one part of a system will have reverberating effects on other parts of the system.

Optimizing the Human Experience

Science, technology, engineering, and mathematics as disciplines have the capacity to continuously improve the ways humans live, interact, and find meaning in the world, thus working to optimize the human experience. This idea has two components: being more suited to our environment and being more fully human. For example, the progression of STEM ideas can help humans create solutions to complex problems, such as improving ways to access water sources, designing energy sources with minimal impact on our environment, developing new ways of communication and expression, and building efficient shelters. STEM ideas can also provide access to the secrets and wonders of nature. Learning in STEM requires students to think logically and systematically, which is a way of knowing the world that is markedly different from knowing the world as an artist. When students can employ various ways of knowing and understand when it is appropriate to use a different way of knowing or integrate ways of knowing, they are fully experiencing the best of what it is to be human. The problem-based learning scenarios provided in the STEM Road Map help students develop ways of thinking like STEM professionals as they ask questions and design solutions. They learn to optimize the human experience by innovating improvements in the designed world in which they live.

THE NEED FOR AN INTEGRATED STEM APPROACH

At a basic level, STEM stands for science, technology, engineering, and mathematics. Over the past decade, however, STEM has evolved to have a much broader scope and implications. Now, educators and policy makers refer to STEM as not only a concentrated area for investing in the future of the United States and other nations but also as a domain and mechanism for educational reform.

The good intentions of the recent decade-plus of focus on accountability and increased testing has resulted in significant decreases not only in instructional time for teaching science and social studies but also in the flexibility of teachers to promote authentic, problem solving–focused classroom environments. The shift has had a detrimental impact on student acquisition of vitally important skills, which many refer to as 21st century skills, and often the ability of students to "think." Further, schooling has become increasingly siloed into compartments of mathematics, science, English language arts, and social studies, lacking any of the connections that are overwhelmingly present in

the real world around children. Students have experienced school as content provided in boxes that must be memorized, devoid of any real-world context, and often have little understanding of why they are learning these things.

STEM-focused projects, curriculum, activities, and schools have emerged as a means to address these challenges. However, most of these efforts have continued to focus on the individual STEM disciplines (predominantly science and engineering) through more STEM classes and after-school programs in a "STEM enhanced" approach (Breiner et al. 2012). But in traditional and STEM enhanced approaches, there is little to no focus on other disciplines that are integral to the context of STEM in the real world. Integrated STEM education, on the other hand, infuses the learning of important STEM content and concepts with a much-needed emphasis on 21st century skills and a problem- and project-based pedagogy that more closely mirrors the real-world setting for society's challenges. It incorporates social studies, English language arts, and the arts as pivotal and necessary (Johnson 2013; Rennie, Venville, and Wallace 2012; Roehrig et al. 2012).

FRAMEWORK FOR STEM INTEGRATION IN THE CLASSROOM

The *STEM Road Map Curriculum Series* is grounded in the Framework for STEM Integration in the Classroom as conceptualized by Moore, Guzey, and Brown (2014) and Moore et al. (2014). The framework has six elements, described in the context of how they are used in the *STEM Road Map Curriculum Series* as follows:

1. The STEM Road Map contexts are meaningful to students and provide motivation to engage with the content. Together, these allow students to have different ways to enter into the challenge.

2. The STEM Road Map modules include engineering design that allows students to design technologies (i.e., products that are part of the designed world) for a compelling purpose.

3. The STEM Road Map modules provide students with the opportunities to learn from failure and redesign based on the lessons learned.

4. The STEM Road Map modules include standards-based disciplinary content as the learning objectives.

5. The STEM Road Map modules include student-centered pedagogies that allow students to grapple with the content, tie their ideas to the context, and learn to think for themselves as they deepen their conceptual knowledge.

6. The STEM Road Map modules emphasize 21st century skills and, in particular, highlight communication and teamwork.

All of the STEM Road Map modules incorporate these six elements; however, the level of emphasis on each of these elements varies based on the challenge or problem in each module.

THE NEED FOR THE *STEM ROAD MAP CURRICULUM SERIES*

As focus is increasing on integrated STEM, and additional schools and programs decide to move their curriculum and instruction in this direction, there is a need for high-quality, research-based curriculum designed with integrated STEM at the core. Several good resources are available to help teachers infuse engineering or more STEM enhanced approaches, but no curriculum exists that spans K–12 with an integrated STEM focus. The next chapter provides detailed information about the specific pedagogy, instructional strategies, and learning theory on which the *STEM Road Map Curriculum Series* is grounded.

REFERENCES

Breiner, J., M. Harkness, C. C. Johnson, and C. Koehler. 2012. What is STEM? A discussion about conceptions of STEM in education and partnerships. *School Science and Mathematics* 112 (1): 3–11.

Johnson, C. C. 2013. Conceptualizing integrated STEM education: Editorial. *School Science and Mathematics* 113 (8): 367–368.

Koehler, C. M., M. A. Bloom, and I. C. Binns. 2013. Lights, camera, action: Developing a methodology to document mainstream films' portrayal of nature of science and scientific inquiry. *Electronic Journal of Science Education* 17 (2).

Moore, T. J., S. S. Guzey, and A. Brown. 2014. Greenhouse design to increase habitable land: An engineering unit. *Science Scope* 51–57.

Moore, T. J., M. S. Stohlmann, H.-H. Wang, K. M. Tank, A. W. Glancy, and G. H. Roehrig. 2014. Implementation and integration of engineering in K–12 STEM education. In *Engineering in pre-college settings: Synthesizing research, policy, and practices*, ed. S. Purzer, J. Strobel, and M. Cardella, 35–60. West Lafayette, IN: Purdue Press.

Rennie, L., G. Venville, and J. Wallace. 2012. *Integrating science, technology, engineering, and mathematics: Issues, reflections, and ways forward.* New York: Routledge.

Roehrig, G. H., T. J. Moore, H. H. Wang, and M. S. Park. 2012. Is adding the *E* enough? Investigating the impact of K–12 engineering standards on the implementation of STEM integration. *School Science and Mathematics* 112 (1): 31–44.

STRATEGIES USED IN THE *STEM ROAD MAP CURRICULUM SERIES*

Erin Peters-Burton, Carla C. Johnson, Toni A. Sondergeld, and Tamara J. Moore

The *STEM Road Map Curriculum Series* uses what has been identified through research as best-practice pedagogy, including embedded formative assessment strategies throughout each module. This chapter briefly describes the key strategies that are employed in the series.

PROJECT- AND PROBLEM-BASED LEARNING

Each module in the *STEM Road Map Curriculum Series* uses either project-based learning or problem-based learning to drive the instruction. Project-based learning begins with a driving question to guide student teams in addressing a contextualized local or community problem or issue. The outcome of project-based instruction is a product that is conceptualized, designed, and tested through a series of scaffolded learning experiences (Blumenfeld et al. 1991; Krajcik and Blumenfeld 2006). Problem-based learning is often grounded in a fictitious scenario, challenge, or problem (Barell 2006; Lambros 2004). On the first day of instruction within the unit, student teams are provided with the context of the problem. Teams work through a series of activities and use open-ended research to develop their potential solution to the problem or challenge, which need not be a tangible product (Johnson 2003).

ENGINEERING DESIGN PROCESS

The *STEM Road Map Curriculum Series* uses engineering design as a way to facilitate integrated STEM within the modules. The engineering design process (EDP) is depicted in Figure 2.1 (p. 10). It highlights two major aspects of engineering design—problem scoping and solution generation—and six specific components of working toward a design: define the problem, learn about the problem, plan a solution, try the solution, test the solution, decide whether the solution is good enough. It also shows that communication

Figure 2.1. Engineering Design Process

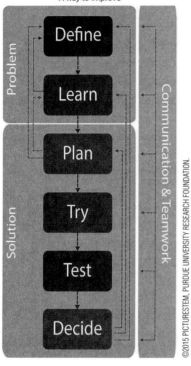

and teamwork are involved throughout the entire process. As the arrows in the figure indicate, the order in which the components of engineering design are addressed depends on what becomes needed as designers progress through the EDP. Designers must communicate and work in teams throughout the process. The EDP is iterative, meaning that components of the process can be repeated as needed until the design is good enough to present to the client as a potential solution to the problem.

Problem scoping is the process of gathering and analyzing information to deeply understand the engineering design problem. It includes defining the problem and learning about the problem. Defining the problem includes identifying the problem, the client, and the end user of the design. The client is the person (or people) who hired the designers to do the work, and the end user is the person (or people) who will use the final design. The designers must also identify the criteria and the constraints of the problem. The criteria are the things the client wants from the solution, and the constraints are the things that limit the possible solutions. The designers must spend significant time learning about the problem, which can include activities such as the following:

- Reading informational texts and researching about relevant concepts or contexts

- Identifying and learning about needed mathematical and scientific skills, knowledge, and tools

- Learning about things done previously to solve similar problems

- Experimenting with possible materials that could be used in the design

Problem scoping also allows designers to consider how to measure the success of the design in addressing specific criteria and staying within the constraints over multiple iterations of solution generation.

Solution generation includes planning a solution, trying the solution, testing the solution, and deciding whether the solution is good enough. Planning the solution includes generating many design ideas that both address the criteria and meet the constraints. Here the designers must consider what was learned about the problem during problem scoping. Design plans include clear communication of design ideas through media such as notebooks, blueprints, schematics, or storyboards. They also include details about the design, such as measurements, materials, colors, costs of materials, instructions for how things fit together, and sets of directions. Making the decision about which design idea to move forward involves considering the trade-offs of each design idea.

Once a clear design plan is in place, the designers must try the solution. Trying the solution includes developing a prototype (a testable model) based on the plan generated. The prototype might be something physical or a process to accomplish a goal. This component of design requires that the designers consider the risk involved in implementing the design. The prototype developed must be tested. Testing the solution includes conducting fair tests that verify whether the plan is a solution that is good enough to meet the client and end user needs and wants. Data need to be collected about the results of the tests of the prototype, and these data should be used to make evidence-based decisions regarding the design choices made in the plan. Here, the designers must again consider the criteria and constraints for the problem.

Using the data gathered from the testing, the designers must decide whether the solution is good enough to meet the client and end user needs and wants by assessment based on the criteria and constraints. Here, the designers must justify or reject design decisions based on the background research gathered while learning about the problem and on the evidence gathered during the testing of the solution. The designers must now decide whether to present the current solution to the client as a possibility or to do more iterations of design on the solution. If they decide that improvements need to be made to the solution, the designers must decide if there is more that needs to be understood about the problem, client, or end user; if another design idea should be tried; or if more planning needs to be conducted on the same design. One way or another, more work needs to be done.

Throughout the process of designing a solution to meet a client's needs and wants, designers work in teams and must communicate to each other, the client, and likely the end user. Teamwork is important in engineering design because multiple perspectives and differing skills and knowledge are valuable when working to solve problems. Communication is key to the success of the designed solution. Designers must communicate their ideas clearly using many different representations, such as text in an engineering notebook, diagrams, flowcharts, technical briefs, or memos to the client.

LEARNING CYCLE

The same format for the learning cycle is used in all grade levels throughout the STEM Road Map, so that students engage in a variety of activities to learn about phenomena in the modules thoroughly and have consistent experiences in the problem- and project-based learning modules. Expectations for learning by younger students are not as high as for older students, but the format of the progression of learning is the same. Students who have learned with curriculum from the STEM Road Map in early grades know

what to expect in later grades. The learning cycle consists of five parts—Introductory Activity/Engagement, Activity/Exploration, Explanation, Elaboration/Application of Knowledge, and Evaluation/Assessment—and is based on the empirically tested 5E model from BSCS (Bybee et al. 2006).

In the Introductory Activity/Engagement phase, teachers introduce the module challenge and use a unique approach designed to pique students' curiosity. This phase gets students to start thinking about what they already know about the topic and begin wondering about key ideas. The Introductory Activity/Engagement phase positions students to be confident about what they are about to learn, because they have prior knowledge, and clues them into what they don't yet know.

In the Activity/Exploration phase, the teacher sets up activities in which students experience a deeper look at the topics that were introduced earlier. Students engage in the activities and generate new questions or consider possibilities using preliminary investigations. Students work independently, in small groups, and in whole-group settings to conduct investigations, resulting in common experiences about the topic and skills involved in the real-world activities. Teachers can assess students' development of concepts and skills based on the common experiences during this phase.

During the Explanation phase, teachers direct students' attention to concepts they need to understand and skills they need to possess to accomplish the challenge. Students participate in activities to demonstrate their knowledge and skills to this point, and teachers can pinpoint gaps in student knowledge during this phase.

In the Elaboration/Application of Knowledge phase, teachers present students with activities that engage in higher-order thinking to create depth and breadth of student knowledge, while connecting ideas across topics within and across STEM. Students apply what they have learned thus far in the module to a new context or elaborate on what they have learned about the topic to a deeper level of detail.

In the last phase, Assessment, teachers give students summative feedback on their knowledge and skills as demonstrated through the challenge. This is not the only point of assessment (as discussed in the section on Embedded Formative Assessments), but it is an assessment of the culmination of the knowledge and skills for the module. Students demonstrate their cognitive growth at this point and reflect on how far they have come since the beginning of the module. The challenges are designed to be multidimensional in the ways students must collaborate and communicate their new knowledge.

STEM RESEARCH NOTEBOOK

One of the main components of the *STEM Road Map Curriculum Series* is the STEM Research Notebook, a place for students to capture their ideas, questions, observations, reflections, evidence of progress, and other items associated with their daily work. At the beginning of each module, the teacher walks students through the setup of the STEM

Research Notebook, which could be a three-ring binder, composition book, or spiral notebook. You may wish to have students create divided sections so that they can easily access work from various disciplines during the module. Electronic notebooks kept on student devices are also acceptable and encouraged. Students will develop their own table of contents and create chapters in the notebook for each module.

Each lesson in the *STEM Road Map Curriculum Series* includes one or more prompts that are designed for inclusion in the STEM Research Notebook and appear as questions or statements that the teacher assigns to students. These prompts require students to apply what they have learned across the lesson to solve the big problem or challenge for that module. Each lesson is designed to meaningfully refer students to the larger problem or challenge they have been assigned to solve with their teams. The STEM Research Notebook is designed to be a key formative assessment tool, as students' daily entries provide evidence of what they are learning. The notebook can be used as a mechanism for dialogue between the teacher and students, as well as for peer and self-evaluation.

The use of the STEM Research Notebook is designed to scaffold student notebooking skills across the grade bands in the *STEM Road Map Curriculum Series*. In the early grades, children learn how to organize their daily work in the notebook as a way to collect their products for future reference. In elementary school, students structure their notebooks to integrate background research along with their daily work and lesson prompts. In the upper grades (middle and high school), students expand their use of research and data gathering through team discussions to more closely mirror the work of STEM experts in the real world.

THE ROLE OF ASSESSMENT IN THE *STEM ROAD MAP CURRICULUM SERIES*

Starting in the middle years and continuing into secondary education, the word *assessment* typically brings grades to mind. These grades may take the form of a letter or a percentage, but they typically are used as a representation of a student's content mastery. If well thought out and implemented, however, classroom assessment can offer teachers, parents, and students valuable information about student learning and misconceptions that does not necessarily come in the form of a grade (Popham 2013).

The *STEM Road Map Curriculum Series* provides a set of assessments for each module. Teachers are encouraged to use assessment information for more than just assigning grades to students. Instead, assessments of activities requiring students to actively engage in their learning, such as student journaling in STEM Research Notebooks, collaborative presentations, and constructing graphic organizers, should be used to move student learning forward. Whereas other curriculum with assessments may include objective-type (multiple-choice or matching) tests, quizzes, or worksheets, we have intentionally avoided these forms of assessments to better align assessment strategies with teacher instruction and

student learning techniques. Since the focus of this book is on project- or problem-based STEM curriculum and instruction that focuses on higher-level thinking skills, appropriate and authentic performance assessments were developed to elicit the most reliable and valid indication of growth in student abilities (Brookhart and Nitko 2008).

Comprehensive Assessment System

Assessment throughout all STEM Road Map curriculum modules acts as a comprehensive system in which formative and summative assessments work together to provide teachers with high-quality information on student learning. Formative assessment occurs when the teacher finds out formally or informally what a student knows about a smaller, defined concept or skill and provides timely feedback to the student about his or her level of proficiency. Summative assessments occur when students have performed all activities in the module and are given a cumulative performance evaluation in which they demonstrate their growth in learning.

A comprehensive assessment system can be thought of as akin to a sporting event. Formative assessments are the practices: It is important to accomplish them consistently, they provide feedback to help students improve their learning, and making mistakes can be worthwhile if students are given an opportunity to learn from them. Summative assessments are the competitions: Students need to be prepared to perform at the best of their ability. Without multiple opportunities to practice skills along the way through formative assessments, students will not have the best chance of demonstrating growth in abilities through summative assessments (Black and Wiliam 1998).

Embedded Formative Assessments

Formative assessments in this module serve two main purposes: to provide feedback to students about their learning and to provide important information for the teacher to inform immediate instructional needs. Providing feedback to students is particularly important when conducting problem- or project-based learning because students take on much of the responsibility for learning, and teachers must facilitate student learning in an informed way. For example, if students are required to conduct research for the Activity/Exploration phase but are not familiar with what constitutes a reliable resource, they may develop misconceptions based on poor information. When a teacher monitors this learning through formative assessments and provides specific feedback related to the instructional goals, students are less likely to develop incomplete or incorrect conceptions in their independent investigations. By using formative assessment to detect problems in student learning and then acting on this information, teachers help move student learning forward through these teachable moments.

Formative assessments come in a variety of formats. They can be informal, such as asking students probing questions related to student knowledge or tasks or simply observing students engaged in an activity to gather information about student skills. Formative assessments can also be formal, such as a written quiz or a laboratory practical. Regardless of the type, three key steps must be completed when using formative assessments (Sondergeld, Bell, and Leusner 2010). First, the assessment is delivered to students so that teachers can collect data. Next, teachers analyze the data (student responses) to determine student strengths and areas that need additional support. Finally, teachers use the results from information collected to modify lessons and create learning environments that reinforce weak points in student learning. If student learning information is not used to modify instruction, the assessment cannot be considered formative in nature.

Formative assessments can be about content, science process skills, or even learning skills. When a formative assessment focuses on content, it assesses student knowledge about the disciplinary core ideas from the *Next Generation Science Standards* (*NGSS*) or content objectives from *Common Core State Standards for Mathematics* (*CCSS Mathematics*) or *Common Core State Standards for English Language Arts* (*CCSS ELA*). Content-focused formative assessments ask students questions about declarative knowledge regarding the concepts they have been learning. Process skills formative assessments examine the extent to which a student can perform science and engineering practices from the *NGSS* or process objectives from *CCSS Mathematics* or *CCSS ELA*, such as constructing an argument. Learning skills can also be assessed formatively by asking students to reflect on the ways they learn best during a module and identify ways they could have learned more.

Assessment Maps

Assessment maps or blueprints can be used to ensure alignment between classroom instruction and assessment. If what students are learning in the classroom is not the same as the content on which they are assessed, the resultant judgment made on student learning will be invalid (Brookhart and Nitko 2008). Therefore, the issue of instruction and assessment alignment is critical. The assessment map for this book (found in Chapter 3) indicates by lesson whether the assessment should be completed as a group or on an individual basis, identifies the assessment as formative or summative in nature, and aligns the assessment with its corresponding learning objectives.

Note that the module includes far more formative assessments than summative assessments. This is done intentionally to provide students with multiple opportunities to practice their learning of new skills before completing a summative assessment. Note also that formative assessments are used to collect information on only one or two learning objectives at a time so that potential relearning or instructional modifications can focus on smaller and more manageable chunks of information. Conversely, summative assessments in the module cover many more learning objectives, as they are traditionally used as final markers of student learning. This is not to say that information collected from summative assessments cannot or should not be used formatively. If teachers find that gaps in student

learning persist after a summative assessment is completed, it is important to revisit these existing misconceptions or areas of weakness before moving on (Black et al. 2003).

SELF-REGULATED LEARNING THEORY IN THE STEM ROAD MAP MODULES

Many learning theories are compatible with the STEM Road Map modules, such as constructivism, situated cognition, and meaningful learning. However, we feel that the self-regulated learning theory (SRL) aligns most appropriately (Zimmerman 2000). SRL requires students to understand that thinking needs to be motivated and managed (Ritchhart, Church, and Morrison 2011). The STEM Road Map modules are student centered and are designed to provide students with choices, concrete hands-on experiences, and opportunities to see and make connections, especially across subjects (Eliason and Jenkins 2012; NAEYC 2016). Additionally, SRL is compatible with the modules because it fosters a learning environment that supports students' motivation, enables students to become aware of their own learning strategies, and requires reflection on learning while experiencing the module (Peters and Kitsantas 2010).

The theory behind SRL (see Figure 2.2) explains the different processes that students engage in before, during, and after a learning task. Because SRL is a cyclical learning process, the accomplishment of one cycle develops strategies for the next learning cycle. This cyclic way of learning aligns with the various sections in the STEM Road Map lesson plans on Introductory Activity/Engagement, Activity/Exploration, Explanation, Elaboration/Application of Knowledge, and Evaluation/Assessment. Since the students engaged in a module take on much of the responsibility for learning, this theory also provides guidance for teachers to keep students on the right track.

The remainder of this section explains how SRL theory is embedded within the five sections of each module and points out ways to

Figure 2.2. SRL Theory

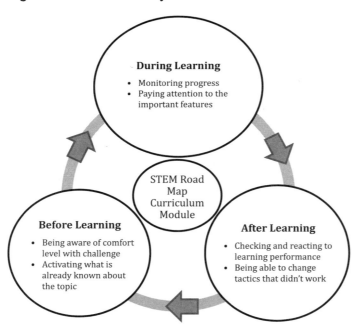

Source: Adapted from Zimmerman 2000.

support students in becoming independent learners of STEM while productively functioning in collaborative teams.

Before Learning: Setting the Stage

Before attempting a learning task such as the STEM Road Map modules, teachers should develop an understanding of their students' level of comfort with the process of accomplishing the learning and determine what they already know about the topic. When students are comfortable with attempting a learning task, they tend to take more risks in learning and as a result achieve deeper learning (Bandura 1986).

The STEM Road Map curriculum modules are designed to foster excitement from the very beginning. Each module has an Introductory Activity/Engagement section that introduces the overall topic from a unique and exciting perspective, engaging the students to learn more so that they can accomplish the challenge. The Introductory Activity also has a design component that helps teachers assess what students already know about the topic of the module. In addition to the deliberate designs in the lesson plans to support SRL, teachers can support a high level of student comfort with the learning challenge by finding out if students have ever accomplished the same kind of task and, if so, asking them to share what worked well for them.

During Learning: Staying the Course

Some students fear inquiry learning because they aren't sure what to do to be successful (Peters 2010). However, the STEM Road Map curriculum modules are embedded with tools to help students pay attention to knowledge and skills that are important for the learning task and to check student understanding along the way. One of the most important processes for learning is the ability for learners to monitor their own progress while performing a learning task (Peters 2012). The modules allow students to monitor their progress with tools such as the STEM Research Notebooks, in which they record what they know and can check whether they have acquired a complete set of knowledge and skills. The STEM Road Map modules support inquiry strategies that include previewing, questioning, predicting, clarifying, observing, discussing, and journaling (Morrison and Milner 2014). Through the use of technology throughout the modules, inquiry is supported by providing students access to resources and data while enabling them to process information, report the findings, collaborate, and develop 21st century skills.

It is important for teachers to encourage students to have an open mind about alternative solutions and procedures (Milner and Sondergeld 2015) when working through the STEM Road Map curriculum modules. Novice learners can have difficulty knowing what to pay attention to and tend to treat each possible avenue for information as equal (Benner 1984). Teachers are the mentors in a classroom and can point out ways for students to approach learning during the Activity/Exploration, Explanation, and

Elaboration/Application of Knowledge portions of the lesson plans to ensure that students pay attention to the important concepts and skills throughout the module. For example, if a student is to demonstrate conceptual awareness of motion when working on roller coaster research, but the student has misconceptions about motion, the teacher can step in and redirect student learning.

After Learning: Knowing What Works

The classroom is a busy place, and it may often seem that there is no time for self-reflection on learning. Although skipping this reflective process may save time in the short term, it reduces the ability to take into account things that worked well and things that didn't so that teaching the module may be improved next time. In the long run, SRL skills are critical for students to become independent learners who can adapt to new situations. By investing the time it takes to teach students SRL skills, teachers can save time later, because students will be able to apply methods and approaches for learning that they have found effective to new situations. In the Evaluation/Assessment portion of the STEM Road Map curriculum modules, as well as in the formative assessments throughout the modules, two processes in the after-learning phase are supported: evaluating one's own performance and accounting for ways to adapt tactics that didn't work well. Students have many opportunities to self-assess in formative assessments, both in groups and individually, using the rubrics provided in the modules.

The designs of the *NGSS* and *CCSS* allow for students to learn in diverse ways, and the STEM Road Map curriculum modules emphasize that students can use a variety of tactics to complete the learning process. For example, students can use STEM Research Notebooks to record what they have learned during the various research activities. Notebook entries might include putting objectives in students' own words, compiling their prior learning on the topic, documenting new learning, providing proof of what they learned, and reflecting on what they felt successful doing and what they felt they still needed to work on. Perhaps students didn't realize that they were supposed to connect what they already knew with what they learned. They could record this and would be prepared in the next learning task to begin connecting prior learning with new learning.

SAFETY IN STEM

Student safety is a primary consideration in all subjects but is an area of particular concern in science, where students may interact with unfamiliar tools and materials that may pose additional safety risks. It is important to implement safety practices within the context of STEM investigations, whether in a classroom laboratory or in the field. When you keep safety in mind as a teacher, you avoid many potential issues with the lesson while also protecting your students.

STEM safety practices encompass things considered in the typical science classroom. Ensure that students are familiar with basic safety considerations, such as wearing protective equipment (e.g., safety glasses or goggles and latex-free gloves) and taking care with sharp objects, and know emergency exit procedures. Teachers should learn beforehand the locations of the safety eyewash, fume hood, fire extinguishers, and emergency shut-off switch in the classroom and how to use them. Also be aware of any school or district safety policies that are in place and apply those that align with the work being conducted in the lesson. It is important to review all safety procedures annually.

STEM investigations should always be supervised. Each lesson in the modules includes teacher guidelines for applicable safety procedures that should be followed. Before each investigation, teachers should go over these safety procedures with the student teams. Some STEM focus areas such as engineering require that students can demonstrate how to properly use equipment in the maker space before the teacher allows them to proceed with the lesson.

Information about classroom science safety, including a safety checklist for science classrooms, general lab safety recommendations, and links to other science safety resources, is available at the Council of State Science Supervisors (CSSS) website at *www.csss-science.org/safety.shtml*. The National Science Teachers Association (NSTA) provides a list of science rules and regulations, including standard operating procedures for lab safety, and a safety acknowledgement form for students and parents or guardians to sign. You can access this forum at *http://static.nsta.org/pdfs/SafetyInTheScienceClassroomLabAndField.pdf*. In addition, NSTA's Safety in the Science Classroom web page (*www.nsta.org/safety*) has numerous links to safety resources, including papers written by the NSTA Safety Advisory Board.

Disclaimer: The safety precautions for each activity are based on use of the recommended materials and instructions, legal safety standards, and better professional practices. Using alternative materials or procedures for these activities may jeopardize the level of safety and therefore is at the user's own risk. Further information regarding safety procedures can be found in other NSTA publications, such as the guide "Safety in the Science Classroom, Laboratory, or Field" (*http://static.nsta.org/pdfs/SafetyInTheScience Classroom.pdf*).

REFERENCES

Bandura, A. 1986. *Social foundations of thought and action: A social cognitive theory.* Englewood Cliffs, NJ: Prentice-Hall.

Barell, J. 2006. *Problem-based learning: An inquiry approach.* Thousand Oaks, CA: Corwin Press.

Benner, P. 1984. *From novice to expert: Excellence and power in clinical nursing practice.* Menlo Park, CA: Addison-Wesley Publishing Company.

Black, P., C. Harrison, C. Lee, B. Marshall, and D. Wiliam. 2003. *Assessment for learning: Putting it into practice.* Berkshire, UK: Open University Press.

Black, P., and D. Wiliam. 1998. Inside the black box: Raising standards through classroom assessment. *Phi Delta Kappan* 80 (2): 139–148.

Blumenfeld, P., E. Soloway, R. Marx, J. Krajcik, M. Guzdial, and A. Palincsar. 1991. Motivating project-based learning: Sustaining the doing, supporting learning. *Educational Psychologist* 26 (3): 369–398.

Brookhart, S. M., and A.J. Nitko. 2008. *Assessment and grading in classrooms.* Upper Saddle River, NJ: Pearson.

Bybee, R., J. Taylor, A. Gardner, P. Van Scotter, J. Carlson, A. Westbrook, and N. Landes. 2006. *The BSCS 5E instructional model: Origins and effectiveness. http://science.education.nih.gov/houseofreps. nsf/b82d55fa138783c2852572c9004f5566/$FILE/Appendix?D.pdf.*

Eliason, C. F., and L. T. Jenkins. 2012. *A practical guide to early childhood curriculum.* 9th ed. New York: Merrill.

Johnson, C. 2003. Bioterrorism is real-world science: Inquiry-based simulation mirrors real life. *Science Scope* 27 (3): 19–23.

Krajcik, J., and P. Blumenfeld. 2006. Project-based learning. In *The Cambridge handbook of the learning sciences,* ed. R. Keith Sawyer, 317–334. New York: Cambridge University Press.

Lambros, A. 2004. *Problem-based learning in middle and high school classrooms: A teacher's guide to implementation.* Thousand Oaks, CA: Corwin Press.

Milner, A. R., and T. Sondergeld. 2015. Gifted urban middle school students: The inquiry continuum and the nature of science. *National Journal of Urban Education and Practice* 8 (3): 442–461.

Morrison, V., and A. R. Milner. 2014. Literacy in support of science: A closer look at cross-curricular instructional practice. *Michigan Reading Journal* 46 (2): 42–56.

National Association for the Education of Young Children (NAEYC). 2016. Developmentally appropriate practice position statements. *www.naeyc.org/positionstatements/dap.*

Peters, E. E. 2010. Shifting to a student-centered science classroom: An exploration of teacher and student changes in perceptions and practices. *Journal of Science Teacher Education* 21 (3): 329–349.

Peters, E. E. 2012. Developing content knowledge in students through explicit teaching of the nature of science: Influences of goal setting and self-monitoring. *Science and Education* 21 (6): 881–898.

Peters, E. E., and A. Kitsantas. 2010. The effect of nature of science metacognitive prompts on science students' content and nature of science knowledge, metacognition, and self-regulatory efficacy. *School Science and Mathematics* 110: 382–396.

Popham, W. J. 2013. *Classroom assessment: What teachers need to know.* 7th ed. Upper Saddle River, NJ: Pearson.

Ritchhart, R., M. Church, and K. Morrison. 2011. *Making thinking visible: How to promote engagement, understanding, and independence for all learners.* San Francisco, CA: Jossey-Bass.

Sondergeld, T. A., C. A. Bell, and D. M. Leusner. 2010. Understanding how teachers engage in formative assessment. *Teaching and Learning* 24 (2): 72–86.

Zimmerman, B. J. 2000. Attaining self-regulation: A social-cognitive perspective. In *Handbook of self-regulation,* ed. M. Boekaerts, P. Pintrich, and M. Zeidner, 13–39. San Diego: Academic Press.

PART 2

WIND ENERGY
STEM ROAD MAP MODULE

WIND ENERGY MODULE OVERVIEW

Janet B. Walton, Sandy Watkins, Carla C. Johnson, and Erin Peters-Burton

THEME: Innovation and Progress

LEAD DISCIPLINES: Social Studies and Science

MODULE SUMMARY

This module focuses on the interactions of Earth's systems, including geography and weather, as well as wind as an energy source. As a culminating activity, students are challenged to develop a proposal for the location of a wind farm. Students investigate U.S. geography, weather patterns, the economics of wind energy, and issues surrounding the use of wind turbines. In science, students learn about how Earth's spheres—the lithosphere, hydrosphere, atmosphere, and biosphere—interact and how these interactions can be observed and measured. Student teams analyze the wind energy potential of their proposed wind farm locations, and each team creates a proposal taking into account factors such as cost, energy production, and environmental impacts of wind farms. Using their findings, student teams deliver presentations on their proposed wind farms with the goal of garnering support from members of the surrounding community and potential investors (adapted from Capobianco et al. 2015).

ESTABLISHED GOALS AND OBJECTIVES

At the conclusion of this module, students will be able to do the following:

- Demonstrate an understanding of the climatic basis of wind

- Identify and describe each of Earth's four spheres: atmosphere, biosphere, hydrosphere, and lithosphere

- Discuss interactions among Earth's spheres and the effects of those interactions

- Identify and discuss the differences between renewable and nonrenewable energy sources

- Demonstrate an understanding of how wind is converted into energy by creating model wind turbines, and identify the major components of a wind turbine

- Identify and describe the steps of the scientific method

- Describe the Engineering Design Process (EDP), which engineers use to design and build products and solve problems

- Apply their understanding of the EDP and the scientific process to plan an investigation, address design challenges, and solve problems

- Apply map skills and their understanding of U.S. geography to identify desirable wind farm sites

- Discuss the economic and environmental implications of wind as an energy source

- Apply their knowledge of the environmental impacts of wind farms to create plans to mitigate those impacts

- Apply their knowledge of economic impacts of wind farms on communities to create a persuasive argument for a wind farm location

CHALLENGE OR PROBLEM FOR STUDENTS TO SOLVE: THE WIND FARM CHALLENGE

As a culminating activity for the module, students participate in the Wind Farm Challenge. In this challenge, students act as energy entrepreneurs who believe that wind energy is the energy of the future. They will choose a location to build a new wind farm and will also try to convince local community members and potential investors that the wind farm is a good idea.

CONTENT STANDARDS ADDRESSED IN THIS STEM ROAD MAP MODULE

A full listing with descriptions of the standards this module addresses can be found in the appendix. Listings of the particular standards addressed within lessons are provided in a table for each lesson in Chapter 4.

STEM RESEARCH NOTEBOOK

Each student should maintain a STEM Research Notebook, which will serve as a place for students to organize their work throughout this module (see p. 12 for more general discussion on setup and use of this notebook). All written work in the module should be included in the notebook, including records of students' thoughts and ideas, fictional accounts based on the concepts in the module, and records of student progress through the EDP. The notebooks may be maintained across subject areas, giving students the opportunity to see that although their classes may be separated during the school day, the knowledge they gain is connected.

Each lesson in this module includes student handouts that should be kept in the STEM Research Notebooks after completion, as well as a prompt to which students should respond in their notebooks. Students will have the opportunity to create covers and tables of contents for their Research Notebooks in Lesson 1. You may also wish to have students include the STEM Research Notebook Guidelines student handout on page 26 in their notebooks.

Emphasize to students the importance of organizing all information in a Research Notebook. Explain to them that scientists and other researchers maintain detailed Research Notebooks in their work. These notebooks, which are crucial to researchers' work because they contain critical information and track the researchers' progress, are often considered legal documents for scientists who are pursuing patents or wish to provide proof of their discovery process.

STUDENT HANDOUT

STEM RESEARCH NOTEBOOK GUIDELINES

STEM professionals record their ideas, inventions, experiments, questions, observations, and other work details in notebooks so that they can use these notebooks to help them think about their projects and the problems they are trying to solve. You will each keep a STEM Research Notebook during this module that is like the notebooks that STEM professionals use. In this notebook, you will include all your work and notes about ideas you have. The notebook will help you connect your daily work with the big problem or challenge you are working to solve.

It is important that you organize your notebook entries under the following headings:

1. **Chapter Topic or Title of Problem or Challenge:** You will start a new chapter in your STEM Research Notebook for each new module. This heading is the topic or title of the big problem or challenge that your team is working to solve in this module.

2. **Date and Topic of Lesson Activity for the Day:** Each day, you will begin your daily entry by writing the date and the day's lesson topic at the top of a new page. Write the page number both on the page and in the table of contents.

3. **Information Gathered From Research:** This is information you find from outside resources such as websites or books.

4. **Information Gained From Class or Discussions With Team Members:** This information includes any notes you take in class and notes about things your team discusses. You can include drawings of your ideas here, too.

5. **New Data Collected From Investigations:** This includes data gathered from experiments, investigations, and activities in class.

6. **Documents:** These are handouts and other resources you may receive in class that will help you solve your big problem or challenge. Paste or staple these documents in your STEM Research Notebook for safekeeping and easy access later.

7. **Personal Reflections:** Here, you record your own thoughts and ideas on what you are learning.

8. **Lesson Prompts:** These are questions or statements that your teacher assigns you within each lesson to help you solve your big problem or challenge. You will respond to the prompts in your notebook.

9. **Other Items:** This section includes any other items your teacher gives you or other ideas or questions you may have.

MODULE LAUNCH

Ask students to share their ideas about where the power for their school and their homes comes from. Create a class list of these ideas. Next, ask students what other sources of energy they can think of, and add these to the list. Ask students if they think that a whole town could be powered by wind. Tell students that there is a town in the United States that gets all its energy from wind. Show a video about the wind energy system of Rock Port, Missouri (visit YouTube and search for "Innovative Cities Rock Port Missouri" or access the video directly at *www.youtube.com/watch?v=mnifNSrZRUQ*). Introduce the module challenge, the Wind Farm Challenge. Tell students that in this challenge they will act as energy entrepreneurs who believe that wind energy is the energy of the future. They will choose a location to build a new wind farm and will also try to convince local community members and potential investors that building it is a good idea. Students will work in teams to design a prototype of a wind turbine, identify an appropriate location for their wind farms, and create plans to mitigate the wind farm's environmental impacts.

PREREQUISITE SKILLS FOR THE MODULE

Students enter this module with a wide range of preexisting skills, information, and knowledge. Table 3.1 (p. 28) provides an overview of prerequisite skills and knowledge that students are expected to apply in this module, along with examples of how they apply this knowledge throughout the module. Differentiation strategies are also provided for students who may need additional support in acquiring or applying this knowledge.

Table 3.1. Prerequisite Key Knowledge and Examples of Applications and Differentiation Strategies

Prerequisite Key Knowledge	Application of Knowledge by Students	Differentiation for Students Needing Additional Knowledge
Measurement skills: • Distance • Time	Measurement skills: • Measure distances, time, and time intervals using standard units.	Measurement skills: • Provide students with opportunities to practice measuring distances using various units and measuring time to the nearest minute. • Provide students with additional content, including textbook support, teacher instruction, and online videos for telling time to the nearest minute.
Map-reading skills	Map-reading skills: • Use U.S. maps to identify regions, landforms, and other geographic features. • Use wind maps to select wind farm sites.	Map-reading skills: • Review basic map-reading skills and knowledge of geography, including continents, oceans, mountains, rivers, canyons, and deserts. • Provide students with map-reading practice.
Inquiry skills: • Ask questions, make logical predictions, plan investigations, and represent data. • Use senses and simple tools to make observations. • Communicate interest in phenomena and plan for simple investigations. • Communicate understanding of simple data using age-appropriate vocabulary.	Inquiry skills: • Select and use appropriate tools and simple equipment to conduct an investigation. • Identify tools needed to investigate specific questions. • Analyze and communicate findings from multiple investigations of similar phenomena to reach a conclusion.	Inquiry skills: • Select model and use appropriate tools and simple equipment to conduct an investigation. • Scaffold student efforts to organize data into appropriate tables, graphs, drawings, or diagrams by providing step-by-step instructions. • Use classroom discussions to identify specific investigations that could be used to answer a particular question and identify reasons for this choice.
Numbers and Operations: • Add and subtract numbers within 1,000. • Multiply and divide whole numbers.	Numbers and Operations: • Engage in activities that involve finding sums of numbers within 1,000. • Understand percentages with a focus on division. • Use division to calculate speed.	Numbers and Operations: • Review and provide models of adding and subtracting within 1,000 using the standard algorithm. • Review multiplication and division and provide examples of calculating percentages.

Table 3.1. (*continued*)

Prerequisite Key Knowledge	Application of Knowledge by Students	Differentiation for Students Needing Additional Knowledge
Reading: • Use information gained from the illustrations and words in a print or digital text to demonstrate understanding of the connection between a series of events, scientific ideas or concepts, or steps in technical procedures in a text.	Reading: • Read informational texts to understand various facets of wind energy, including financial costs, economic benefits, and environmental impacts.	Reading: • Provide reading strategies to support comprehension of nonfiction texts, including activating prior knowledge, previewing text by skimming content and scanning images, and rereading.
Writing: • Write informative/explanatory and persuasive texts in which students introduce a topic, use facts and definitions to develop points, and provide a concluding statement or section.	Writing: • Write informative/explanatory and persuasive texts to examine a topic and convey ideas and information clearly. • Write narratives to develop real or imagined experiences or events using effective technique, descriptive details, and clear event sequences.	Writing: • Provide a template for writing informative/explanatory texts to scaffold student writing exercises. • Provide writing organizer handouts to scaffold student work in describing details and clarifying event sequence.

POTENTIAL STEM MISCONCEPTIONS

Students enter the classroom with a wide variety of prior knowledge and ideas, so it is important to be alert to misconceptions, or inappropriate understandings of foundational knowledge. These misconceptions can be classified as one of several types: "preconceived notions," opinions based on popular beliefs or understandings; "nonscientific beliefs," knowledge students have gained about science from sources outside the scientific community; "conceptual misunderstandings," incorrect conceptual models based on incomplete understanding of concepts; "vernacular misconceptions," misunderstandings of words based on their common use versus their scientific use; and "factual misconceptions," incorrect or imprecise knowledge learned in early life that remains unchallenged (NRC 1997, p. 28). Misconceptions must be addressed and dismantled in order for students to reconstruct their knowledge, and therefore teachers should be prepared to take the following steps:

- *Identify students' misconceptions.*

- *Provide a forum for students to confront their misconceptions.*

- *Help students reconstruct and internalize their knowledge, based on scientific models. (NRC 1997, p. 29)*

Keeley and Harrington (2010) recommend using diagnostic tools such as probes and formative assessment to identify and confront student misconceptions and begin the process of reconstructing student knowledge. Keeley and Harrington's *Uncovering Student Ideas in Science* series contains probes targeted toward uncovering student misconceptions in a variety of areas. In particular, the portions of volume 2 (Keeley and Eberle 2008) and volume 3 (Keeley and Harrington 2014) that address electricity and the nature of science may be useful resources for addressing student misconceptions in this module.

Some commonly held misconceptions specific to lesson content are provided with each lesson so that you can be alert for student misunderstanding of the science concepts presented and used during this module. The American Association for the Advancement of Science has also identified misconceptions that students frequently hold regarding various science concepts (see the links at *http://assessment.aaas.org/topics*).

SRL PROCESS COMPONENTS

Table 3.2 illustrates some of the activities in the Wind Energy module and how they align to the SRL learning processes before, during, and after learning.

Table 3.2. SRL Process Components

Learning Process Components	Example From Wind Energy Module	Lesson Number
BEFORE LEARNING		
Motivates students	Students create a class list of sources of power and energy for their school and homes. Then they learn about one town that is entirely powered by wind.	Lesson 1
Evokes prior learning	Students use their own environment to consider the type of energy that is produced for their own consumption.	Lesson 1
DURING LEARNING		
Focuses on important features	Students explore factors that optimize wind resources and technologies for a particular type of geographic location.	Lesson 2
Helps students monitor their progress	Students share results of Don't Bother the Neighbors activity in presentations. Students observe what other students focus on in their presentations.	Lesson 3
AFTER LEARNING		
Evaluates learning	Students get feedback on their final challenge products from detailed rubrics.	Lesson 4
Takes account of what worked and what did not work	Students respond to questions from the audience after their Wind Farm Challenge presentation.	Lesson 4

STRATEGIES FOR DIFFERENTIATING INSTRUCTION WITHIN THIS MODULE

For the purposes of this curriculum module, differentiated instruction is conceptualized as a way to tailor instruction—including process, content, and product—to various student needs in your class. A number of differentiation strategies are integrated into lessons across the module. The problem- and project-based learning approach used in the lessons is designed to address students' multiple intelligences by providing a variety of entry points and methods to investigate the key concepts in the module. Differentiation

strategies for students needing support in prerequisite knowledge can be found in Table 3.1 (p. 28). You are encouraged to use information gained about student prior knowledge during introductory activities and discussions to inform your instructional differentiation. Strategies incorporated into this lesson include flexible grouping, varied environmental learning contexts, assessments, compacting, and tiered assignments and scaffolding.

Flexible Grouping: Students work collaboratively in a variety of activities throughout this module. Grouping strategies you might employ include student-led grouping, grouping students according to ability level, grouping students randomly, or grouping them so that students in each group have complementary strengths (for instance, one student might be strong in mathematics, another in art, and another in writing). You may also choose to group students based on their prior knowledge. For Lesson 2, you may choose to maintain the same student groupings as in Lesson 1 or regroup students according to another of the strategies described here. You may therefore wish to consider grouping students in Lesson 2 into design teams on which they will remain throughout the module.

Varied Environmental Learning Contexts: Students have the opportunity to learn in various contexts throughout the module, including alone, in groups, in quiet reading and research-oriented activities, and in active learning through inquiry and design activities. In addition, students learn in a variety of ways, including through doing inquiry activities, reading fiction and nonfiction texts, watching videos, participating in class discussion, and conducting web-based research.

Assessments: Students are assessed in a variety of ways throughout the module, including individual and collaborative formative and summative assessments. Students have the opportunity to produce work via written text, oral and media presentations, and modeling. You may choose to provide students with additional choices of media for their products (for example, PowerPoint presentations, posters, or student-created websites or blogs).

Compacting: Based on student prior knowledge, you may wish to adjust instructional activities for students who exhibit prior mastery of a learning objective. For instance, if some students exhibit mastery of calculating wind speeds in Lesson 1, you may wish to limit the amount of time they spend practicing these skills and instead introduce various units of measurement and unit conversions to these students.

Tiered Assignments and Scaffolding: Based on your awareness of student ability, understanding of concepts, and mastery of skills, you may wish to provide students with variations on activities by adding complexity to assignments or providing more or fewer learning supports for activities throughout the module. For instance, some students may need additional support in identifying key search words and phrases for web-based research or may benefit from cloze sentence handouts to enhance vocabulary understanding. Other students may benefit from expanded reading selections and additional reflective writing or from working with manipulatives and other visual representations of mathematical concepts. You may also work with your school librarian to compile a set of topical resources at a variety of reading levels.

STRATEGIES FOR ENGLISH LANGUAGE LEARNERS

Students who are developing proficiency in English language skills require additional supports to simultaneously learn academic content and the specialized language associated with specific content areas. WIDA has created a framework for providing support to these students and makes available rubrics and guidance on differentiating instructional materials for English language learners (ELLs) (see *www.wida.us/get.aspx?id=7*). In particular, ELL students may benefit from additional sensory supports such as images, physical modeling, and graphic representations of module content, as well as interactive support through collaborative work. This module incorporates a variety of sensory supports and offers ongoing opportunities for ELL students to work with collaboratively. The focus in this module on wind energy affords opportunities to access the culturally diverse experiences of ELL students in the classroom because students may have varied geographical backgrounds and diverse experiences with the ways that natural resource availability influences electricity generation (e.g., capturing wind energy and solar power).

Teachers differentiating instruction for ELL students should carefully consider the needs of these students as they introduce and use academic language in various language domains (listening, speaking, reading, and writing) throughout this module. To adequately differentiate instruction for ELL students, teachers should have an understanding of the proficiency level of each student. The following five overarching preK–5 WIDA learning standards are relevant to this module:

- Standard 1: Social and Instructional language. Focus on social behavior in group work and class discussions.

- Standard 2: The language of Language Arts. Focus on forms of print, elements of text, picture books, comprehension strategies, main ideas and details, persuasive language, creation of informational text, and editing and revision.

- Standard 3: The language of Mathematics. Focus on numbers and operations, patterns, number sense, measurement, and strategies for problem solving.

- Standard 4: The language of Science. Focus on safety practices, energy sources, scientific process, and scientific inquiry.

- Standard 5: The language of Social Studies. Focus on change from past to present, historical events, resources, map reading, and location of objects and places.

SAFETY CONSIDERATIONS FOR THE ACTIVITIES IN THIS MODULE

Science activities in this module focus on wind energy. Students create wind turbines from a variety of materials and should use caution when handling materials with sharp edges, particularly as they spin. For more general safety guidelines, see the Safety in STEM section in Chapter 2 (p. 18).

DESIRED OUTCOMES AND MONITORING SUCCESS

The desired outcomes for this module are outlined in Table 3.3, along with suggested ways to gather evidence to monitor student success. For more specific details on desired outcomes, see the Established Goals and Objectives sections for the module and individual lessons.

Table 3.3. Desired Outcomes and Evidence of Success in Achieving Identified Outcomes

Desired Outcomes	Evidence of Success	
	Performance Tasks	Other Measures
Students can apply an understanding of geography, mapping, and various science and mathematics concepts to complete small group projects and individual tasks related to the projects within the module.	• Students maintain STEM Research Notebooks that contain designs, research notes, evidence of collaboration, and ELA-related work. • Student teams design prototypes of wind turbines and create plans to mitigate the environmental impacts of a wind farm. • Student teams research and present information on a wind turbine location. • Students are able to discuss how they applied their understanding of concepts introduced in the unit to their designs (individual and team) and presentations. • Students are assessed using project rubrics that focus on content and application of skills related to the academic content.	Student collaboration is assessed using a collaboration rubric.

ASSESSMENT PLAN OVERVIEW AND MAP

Table 3.4 provides an overview of the major group and individual *products and deliverables,* or things that student teams will produce in this module, that constitute the assessment for this module. See Table 3.5 (p. 36) for a full assessment map of formative and summative assessments in this module.

Table 3.4. Major Products and Deliverables in Lead Disciplines for Groups and Individuals

Lesson	Major Group Products and Deliverables	Major Individual Products and Deliverables
1	• Blown Away Design Challenge boat • Resource Your Day research and handout • Earth's Spheres model/presentation	• Enough for Everyone? handout • Scarcity Scramble handouts • Blown Away Engineer It! handouts • Connect the Spheres handout • Earth's Spheres graphic organizer • STEM Research Notebook prompt
2	• U.S. Map Scavenger Hunt handout • How Windy Is the Wind? anemometer	• Map Me! handouts and maps • Where's the Wind? handouts • How Windy Is the Wind? Engineer It! handouts • Earth's Spheres Assessment • STEM Research Notebook prompt • Evidence of collaboration
3	• Dollars and Wind budget • Don't Bother the Neighbors presentation • Catch the Wind pinwheel • Energy Explorers data and calculations	• Catch the Wind data sheets • Don't Bother the Neighbors Engineer It! handouts • Scientific Method Assessment • STEM Research Notebook prompt • Evidence of collaboration
4	• Wind Farm Challenge presentation and model	• Wind Farm Challenge Economic Benefits handouts • Wind Farm Challenge Careers handouts • Wind Farm Challenge presentation graphic organizers • STEM Research Notebook prompt • Evidence of collaboration

Table 3.5. Assessment Map for Wind Energy Module

Lesson	Assessment	Group/ Individual	Formative/ Summative	Lesson Objective Assessed
1	Blown Away Design Challenge *boat*	Group	Formative	• Understand that engineers use a process, the EDP, to solve problems and create products. • Apply understanding of wind energy and the EDP to a design challenge.
1	Resource Your Day *research*	Group	Formative	• Identify renewable and nonrenewable resources and discuss how these are used in their daily lives. • Describe the power source for their homes and schools.
1	Earth's Spheres *model/ presentation*	Group	Formative	• Understand that wind is the movement of air due to pressure differentials and Earth's rotation. • Understand and discuss the role of Earth's systems in the creation of wind.
1	Enough for Everyone? *handout*	Individual	Formative	• Understand and discuss the concept of resource scarcity.
1	Resource Your Day *handout*	Individual	Formative	• Identify renewable and nonrenewable resources and discuss how these are used in their daily lives
1	Scarcity Scramble *handout*	Individual	Formative	• Identify renewable and nonrenewable resources and discuss how these are used in their daily lives • Define and discuss the concept of resource scarcity

Table 3.5. (*continued*)

Lesson	Assessment	Group/ Individual	Formative/ Summative	Lesson Objective Assessed
1	Blown Away Engineer It! *handouts*	Individual	Formative	• Understand that engineers use a process, the EDP, to solve problems and create products. • Apply understanding of wind energy and the EDP to a design challenge.
1	Connect the Spheres *handout*	Individual	Formative	• Identify Earth's four spheres: atmosphere, biosphere, hydrosphere, and lithosphere • Describe and illustrate interactions among Earth's spheres
1	Earth's Spheres *graphic organizer*	Individual	Formative	• Understand and discuss the role of Earth's systems in the creation of wind.
1	STEM Research Notebook *prompts*	Individual	Formative	• Discuss the role of Earth's spheres in the creation of wind • Understand and discuss the concept of resource scarcity.
2	How Windy Is the Wind? *anemometer*	Group	Formative	• Apply understanding of anemometer design and function and of the EDP to design and build a functioning anemometer.
2	How Windy Is the Wind? Engineer It! *handout*	Individual	Formative	• Apply understanding of anemometer design and function and of the EDP to design and build a functioning anemometer.
2	Map Me! *handouts and map*	Individual	Formative	• Identify geographic features of a U.S. region and create a map of these features. • Apply understanding of map symbols and conventions to locate landforms and geographic locations on maps.

Table 3.5. (*continued*)

Lesson	Assessment	Group/ Individual	Formative/ Summative	Lesson Objective Assessed
2	Where's the Wind? *handouts*	Individual	Formative	• Apply map-reading skills to locate and identify existing wind farms in a region of the United States. • Apply understanding of map symbols and conventions to locate landforms and geographic locations on maps. • Apply map-reading skills to identify average wind speeds in a region of the United States. • Apply understanding of the differences in geography across the United States to make predictions about wind resource availability.
2	Earth's Spheres Assessment	Individual	Summative	• Identify Earth's four spheres: atmosphere, biosphere, hydrosphere, and lithosphere. • Describe and illustrate interactions among Earth's spheres. • Discuss the role of Earth's spheres in the creation of wind. • Create a model that illustrates interactions among Earth's spheres.
2	STEM Research Notebook *prompt*	Individual	Formative	• Identify differences and similarities between sedimentary, igneous, and metamorphic rocks..
2	Evidence of collaboration	Individual	Formative	• Collaborate with peers to create a solution to a problem.

Table 3.5. (*continued*)

Lesson	Assessment	Group/ Individual	Formative/ Summative	Lesson Objective Assessed
3	Dollars and Wind *budget*	Group	Formative	• Research costs and income streams associated with wind farms and apply this information to create a budget.
3	Don't Bother the Neighbors *presentation*	Group	Formative	• Understand that there are environmental impacts associated with wind farms, but that these impacts are substantially different from those associated with burning fossil fuels. • Apply understanding of the environmental impacts of wind farms to create a plan to mitigate one type of environmental disadvantage.
3	Energy Explorers *data and calculations*	Group	Formative	• Apply basic mathematical skills to understand average household energy usage and how various energy sources can meet consumer energy needs.
3	Catch the Wind *data sheets*	Individual	Formative	• Identify the basic parts of a wind turbine. • Apply understanding of wind turbine technology to build a simple wind turbine and measure the amount of electricity it can produce. • Apply understanding of wind turbine technology to create a variety of turbine blades. • Create a plan for a scientific investigation using the scientific method.

Table 3.5. (*continued*)

Lesson	Assessment	Group/Individual	Formative/Summative	Lesson Objective Assessed
3	Don't Bother the Neighbors Engineer It! *handouts*	Individual	Formative	• Apply understanding of the environmental impacts of wind farms to create a plan to mitigate one type of environmental disadvantage. • Use the EDP to design an innovation to mitigate an environmental impact of wind farms.
3	Scientific Method Assessment	Individual	Summative	• Identify tools scientists use to measure natural phenomena in each of Earth's spheres. • Identify and describe the steps of the scientific method. • Develop a testable question and a hypothesis for a scientific investigation.
3	STEM Research Notebook *prompt*	Individual	Formative	• Understand that there are environmental impacts associated with wind farms, but that these impacts are substantially different from those associated with burning fossil fuels.
3	Evidence of collaboration	Individual	Formative	• Collaborate with peers to create a solution to a problem.

Table 3.5. (*continued*)

Lesson	Assessment	Group/ Individual	Formative/ Summative	Lesson Objective Assessed
4	Wind Farm Challenge *presentation and model*	Group	Summative	• Apply understanding of economic, environmental, and technological features of wind turbines and wind farms to create a proposal for a wind farm location. • Identify careers related to the wind energy industry. • Create a persuasive argument for a wind farm location. • Demonstrate an understanding of the basic components and function of a wind turbine. • Collaborate with peers to create a solution to a problem.
4	Wind Farm Challenge individual student *handouts*	Individual	Summative	• Apply understanding of economic, environmental, and technological features of wind turbines and wind farms to create a proposal for a wind farm location. • Identify careers related to the wind energy industry. • Create a persuasive argument for a wind farm location.
4	Evidence of collaboration	Individual	Formative	• Collaborate with peers to create a solution to a problem.
4	STEM Research Notebook *prompt*	Individual	Formative	• Demonstrate an understanding of the basic components and function of a wind turbine.

MODULE TIMELINE

Tables 3.6–3.10 (pp. 42–43) provide lesson timelines for each week of the module. These timelines are provided for general guidance only and are based on class times of approximately 45 minutes.

Table 3.6. STEM Road Map Module Schedule for Week One

Day 1	Day 2	Day 3	Day 4	Day 5
Lesson 1: *The Wonderful Wind*	*Lesson 1:* *The Wonderful Wind*	*Lesson 1:* *The Wonderful Wind*	*Lesson 1:* *The Wonderful Wind*	*Lesson 1:* *The Wonderful Wind*
• Launch the module by introducing nonrenewable and renewable energy sources, wind energy, and the concept of scarcity.	• Investigate scarcity with Enough for Everyone? activity.	• Investigate natural resources and renewable versus nonrenewable resources in Resource Your Day activity.	• Continue Resource Your Day activity.	• Investigate the distribution of resources in Scarcity Scramble activity.
• Introduce STEM Research Notebook.	• Begin Earth's Spheres activity.	• Complete Earth's Spheres activity.	• Apply the EDP in Blown Away Design Challenge.	
• Introduce the Wind Farm Challenge.		• Introduce the EDP.		

Table 3.7. STEM Road Map Module Schedule for Week Two

Day 6	Day 7	Day 8	Day 9	Day 10
Lesson 2: *Where's the Wind?*	*Lesson 2:* *Where's the Wind?*	*Lesson 2:* *Where's the Wind?*	*Lesson 2:* *Where's the Wind?*	*Lesson 2:* *Where's the Wind?*
• Introduce wind turbine technology.	• Introduce maps and the various types available in Marvelous Maps activity.	• Students investigate a region of the United States in Map Me! activity.	• Continue Map Me! activity	• Begin Where's the Wind? activity.
• Discuss wind resources as they relate to geography.	• Begin How Windy Is the Wind? activity.	• Continue How Windy Is the Wind? activity.		
• Investigate hurricane force winds.				

Table 3.8. STEM Road Map Module Schedule for Week Three

Day 11	Day 12	Day 13	Day 14	Day 15
Lesson 2: Where's the Wind? • Continue Where's the Wind? activity	*Lesson 2: Where's the Wind?* • Complete Where's the Wind? activity. • Explore meteorology/weather websites.	*Lesson 3: Wind Impact* • Introduce environmental and financial costs of wind farms.	*Lesson 3: Wind Impact* • Introduce Dollars and Wind activity. • Students investigate wind turbine electricity generation and blade design in Catch the Wind investigation.	*Lesson 3: Wind Impact* • Complete Dollars and Wind activity. • Introduce Don't Bother the Neighbors activity. • Continue Catch the Wind investigation.

Table 3.9. STEM Road Map Module Schedule for Week Four

Day 16	Day 17	Day 18	Day 19	Day 20
Lesson 3: Wind Impact • Continue Don't Bother the Neighbors activity.	*Lesson 3: Wind Impact* • Complete Don't Bother the Neighbors activity. • Student give presentations.	*Lesson 4: The Wind Farm Challenge* • Introduce Wind Farm Challenge. • Review challenge materials. • Students investigate local economic benefits of wind farms and careers in energy.	*Lesson 4: The Wind Farm Challenge* • Complete careers research. • Create model of plan or device to mitigate environmental impact.	*Lesson 4: The Wind Farm Challenge* • Organize information for presentation and create the summary.

Table 3.10. STEM Road Map Module Schedule for Week Five

Day 21	Day 22	Day 23	Day 24	Day 25
Lesson 4: The Wind Farm Challenge • Complete information organization and begin work on creating presentation.	*Lesson 4: The Wind Farm Challenge* • Continue work on creating presentations.	*Lesson 4: The Wind Farm Challenge* • Continue work on creating presentations.	*Lesson 4: The Wind Farm Challenge* • Share presentations.	*Lesson 4: The Wind Farm Challenge* • Continue to share presentations.

RESOURCES

Teachers have the option to co-teach portions of this unit and may want to combine classes for activities such as researching the environmental impacts of wind farms. The media specialist can help teachers locate resources for students to view and read about wind energy and provide technical help. Special educators and reading specialists can help find supplemental sources for students needing extra support in reading and writing. Additional resources may be found online. Community resources for this module may include urban planners, engineers, school administrators, and parents.

REFERENCES

Capobianco, B. M., C. Parker, A. Laurier, and J. Rankin. 2015. The STEM Road Map for grades 3–5. In *STEM Road Map: A framework for integrated STEM education*, ed. C. C. Johnson, E. E. Peters-Burton, and T. J. Moore, 68–95. New York: Routledge. *www.routledge.com/products/9781138804234*.

Keeley, P., and F. Eberle. 2008. *Uncovering student ideas in science, volume 3: Another 25 formative assessment probes.* Arlington, VA: NSTA Press.

Keeley, P., and R. Harrington. 2014. *Uncovering student ideas in physical science, volume 2: 39 new electricity and magnetism formative assessment probes.* Arlington, VA: NSTA Press.

WIDA. 2012. 2012 Amplification of the English language development standards: Kindergarten–grade 12. *www.wida.us/standards/eld.aspx*.

WIND ENERGY LESSON PLANS

Janet B. Walton, Sandy Watkins, Carla C. Johnson, and Erin Peters-Burton

Lesson Plan 1: The Wonderful Wind

In this lesson, students learn the distinction between nonrenewable and renewable sources of energy, recognizing that wind energy is one of the primary renewable sources used today. They also learn about the concept of resource scarcity as it pertains to energy and other natural resources. Students gain a conceptual understanding of how wind is converted into electricity and can therefore power homes, businesses, and schools in a community. In science, students investigate wind as an atmospheric phenomenon in the context of Earth's spheres. Students are introduced to the engineering design process (EDP) as a structure for group problem solving and are challenged to apply the EDP to a design problem.

ESSENTIAL QUESTIONS

- How can wind be used as a renewable source of energy for communities?
- How does resource scarcity affect populations worldwide?
- How do renewable energy sources address scarcity?

ESTABLISHED GOALS AND OBJECTIVES

At the conclusion of this lesson, students will be able to do the following:

- Identify renewable and nonrenewable resources and discuss how these are used in their daily lives
- Describe the power sources for their homes and schools
- Define and discuss the concept of resource scarcity
- Identify Earth's four spheres: atmosphere, biosphere, hydrosphere, and lithosphere
- Describe and illustrate interactions among Earth's spheres
- Discuss the role of Earth's spheres in the creation of wind
- Create a model that illustrates interactions among Earth's spheres

- Identify several limitations of models
- Identify the engineering design process (EDP) as a series of steps that engineers use to solve problems and create products
- Apply their understanding of wind energy and the EDP to create a solution to a design challenge

TIME REQUIRED

- 5 days (approximately 45 minutes each day; see Table 3.6, p. 42)

MATERIALS

Required Materials for Lesson 1

- STEM Research Notebooks (1 per student; see p. 26 for STEM Research Notebook student handout)
- Internet access for student research and viewing videos
- Handouts (attached at the end of this lesson)

Additional Materials for Enough for Everyone? (per group unless otherwise noted)

- 60 small pieces of candy in varied colors, such as M&Ms, Skittles, or jelly beans* (*Note:* If you wish to allow students to eat candy afterward, purchase extra, as students will handle the candy in groups during the activity. Remind students to not eat food used in the lab activity or in the lab.)
- Roll of paper towels (per class)
- 2 paper bags*
- Plastic resealable bag (quart size)
- 2 sets of 15 slips of paper, each set numbered 1 to 15
- Enough for Everyone? handout (1 per student; attached at the end of this lesson)

Additional Materials for Scarcity Scramble

- Envelopes (1 per group)*
- Slips of paper labeled with resource units*
- Scarcity Scramble handouts (1 set per student; attached at the end of this lesson)

*See Preparation for Lesson 1 on page 55 for more details.

Additional Materials for Earth's Spheres

- Connect the Spheres handout (5 copies per student; attached at the end of this lesson)
- Earth's Spheres graphic organizer (1 per student; attached at the end of this lesson)
- Chart paper (1–2 sheets per group of 3–4 students)
- Markers

Additional Materials for Blown Away Design Challenge (per group unless otherwise noted)

- Small plastic or inflatable wading pool (1 per class)
- Water to fill pool a few inches deep
- 1 small plastic container (about 4 × 6 inches)
- Fan (1 per class)
- 5 small craft sticks
- 5 coffee stirrers
- 2 plastic drinking straws
- 3 index cards (3 × 5 inches)
- 1 piece of notebook or printer paper
- 1 piece of tissue paper
- 2 small paper cups (3 or 5 oz. size)
- 1 sheet of aluminum foil (about 12 × 6 inches)
- 1 sheet of waxed paper (about 12 × 6 inches)
- 1 plastic grocery bag
- 1 pair of scissors
- 1 roll of masking tape
- 8 large paper clips
- Blown Away Engineer It! handouts (1 set per student; attached at the end of this lesson)
- Safety glasses or goggles

SAFETY NOTES

1. All laboratory occupants must wear safety glasses or googles during all phases of this inquiry activity.

2. Do not eat any food used during this investigation.

3. Immediately wipe up any water on the floor to avoid a slip-and-fall hazard.

4. Keep away from electrical sources when working with water in the pool because of the shock hazard.

5. Use caution when operating electrical devices (e.g., fans) because of the potential shock hazard, especially near water.

6. Use caution when working with sharps (scissors, sticks, stirrers, fan blades, etc.) to avoid cutting or puncturing skin.

7. Make sure all materials are put away after completing the activity.

8. Wash hands with soap and water after completing the activity.

CONTENT STANDARDS AND KEY VOCABULARY

Table 4.1 lists the content standards from the *Next Generation Science Standards (NGSS)*, *Common Core State Standards,* and the Framework for 21st Century Learning that this lesson addresses, and Table 4.2 (p. 51) presents the key vocabulary. Vocabulary terms are provided for both teacher and student use. Teachers may choose to introduce some or all of the terms to students.

Table 4.1. Content Standards Addressed in STEM Road Map Module Lesson 1

NEXT GENERATION SCIENCE STANDARDS **PERFORMANCE EXPECTATIONS** • 5-ESS2-1. Develop a model using an example to describe ways the geosphere, biosphere, hydrosphere, and/or atmosphere interact. • 3-5-ETS1-1. Define a simple design problem reflecting a need or a want that includes specified criteria for success and constraints on materials, time, or cost. • 3-5-ETS1-2. Generate and compare multiple possible solutions to a problem based on how well each is likely to meet the criteria and constraints of the problem.

Table 4.1. (*continued*)

DISCIPLINARY CORE IDEAS

ESS3.A. Natural Resources
- Energy and fuels that humans use are derived from natural sources, and their use affects the environment in multiple ways. Some resources are renewable over time, and others are not.

ETS1.A. Defining and Delimiting Engineering Problems
- Possible solutions to a problem are limited by available materials and resources (constraints). The success of a designed solution is determined by considering the desired features of a solution (criteria). Different proposals for solutions can be compared on the basis of how well each one meets the specified criteria for success or how well each takes the constraints into account. (3-5-ETS1-1)

CROSSCUTTING CONCEPTS

Systems and System Models
- A system is a group of related parts that make up a whole and can carry out functions its individual parts cannot.
- A system can be described in terms of its components and their interactions.

Energy and Matter
- Energy can be transferred in various ways and between objects.

Influence of Science, Engineering, and Technology on Society and the Natural World
- People's needs and wants change over time, as do their demands for new and improved technologies.
- Engineers improve existing technologies or develop new ones to increase their benefits, decrease known risks, and meet societal demands.

SCIENCE AND ENGINEERING PRACTICES

Asking Questions and Defining Problems
- Use prior knowledge to describe problems that can be solved.
- Define a simple design problem that can be solved through the development of an object, tool, process, or system and includes several criteria for success and constraints on materials, time, or cost.

Developing and Using Models
- Identify limitations of models.
- Collaboratively develop and/or revise a model based on evidence that shows the relationships among variables for frequent and regular occurring events.
- Develop a model using an analogy, example, or abstract representation to describe a scientific principle or design solution.
- Develop and/or use models to describe and/or predict phenomena.

Table 4.1. (*continued*)

Developing and Using Models (*continued*)

- Develop a diagram or simple physical prototype to convey a proposed object, tool, or process.

- Use a model to test cause and effect relationships or interactions concerning the functioning of a natural or designed system.

COMMON CORE STATE STANDARDS FOR MATHEMATICS

MATHEMATICAL PRACTICES

- MP1. Make sense of problems and persevere in solving them.

- MP2. Reason abstractly and quantitatively.

COMMON CORE STATE STANDARDS FOR ENGLISH LANGUAGE ARTS

READING STANDARDS

- RI.5.4. Determine the meaning of general academic and domain-specific words and phrases in a text relevant to a grade 5 topic or subject area.

- RI.5.7. Draw on information from multiple print or digital sources, demonstrating the ability to locate an answer to a question quickly or to solve a problem efficiently.

- RF.5.3. Know and apply grade-level phonics and word analysis skills in decoding words.

- RF.5.4. Read with sufficient accuracy and fluency to support comprehension.

- RF.5.4.a. Read grade-level text.

WRITING STANDARDS

- W.5.2. Write informative/explanatory texts to examine a topic and convey ideas and information clearly.

- W.5.4. Produce clear and coherent writing in which the development and organization are appropriate to task, purpose, and audience.

SPEAKING AND LISTENING STANDARDS

- SL.5.1. Engage effectively in a range of collaborative discussions (one-on-one, in groups, and teacher-led) with diverse partners on grade 5 topics and texts, building on others' ideas and expressing their own clearly.

- SL.5.1.b. Follow agreed-upon rules for discussions and carry out assigned roles.

- SL.5.1d. Review the key ideas expressed and draw conclusions in light of information and knowledge gained from the discussions.

FRAMEWORK FOR 21ST CENTURY LEARNING

Interdisciplinary themes (financial, economic, and business literacy; environmental literacy); Learning and Innovation Skills; Information, Media and Technology Skills; Life and Career Skills

Table 4.2. Key Vocabulary in Lesson 1

Key Vocabulary	Definition
atmosphere	the layer of gases surrounding Earth
biosphere	the outer parts of Earth that are occupied by living things; includes regions of the surface, atmosphere, and hydrosphere
collaboration	the act of working together in groups to achieve a goal or create something
electricity	a form of energy that is carried through wires and provides power to buildings, lights, machines, and other devices
entrepreneur	a person who starts and operates one or more businesses
hydrosphere	all the water on Earth's surface
lithosphere	Earth's solid outer surface, consisting of its crust and mantle
model	a simplified representation of a system or object that contains important features of the system or object
natural resources	substances that occur in nature and that may be used to make other products
nonrenewable energy sources	energy sources that cannot be replaced when used, such as fossil fuels
renewable energy sources	sources of energy that are naturally replenished when used, such as the Sun or wind
resources	a supply of materials that people or groups draw on to meet wants and needs
scarcity	the state of something being in short supply
scientific model	a representation to make a concept or object easier to understand; types of models include visual (such as a flowchart), physical (a globe), conceptual (weather forecasting), and mathematical models
system	a set of parts that interact with one another to create a complex whole
wind	moving air caused by the uneven heating of Earth's surface by the Sun
wind turbine	a tall tower with rotating blades attached at the top that uses wind to create electricity

TEACHER BACKGROUND INFORMATION

This lesson introduces the module and the final challenge by connecting the concept of scarcity to energy sources and other resources we rely on daily. Wind energy is introduced as a viable option to nonrenewable energy sources as a way to address some of the scarcity issues surrounding fossil fuels. For students to more fully understand wind as a resource with variable availability, they investigate the origins of wind by considering the interaction of Earth's spheres. Students use both the EDP and the scientific method throughout the module, and the EDP is introduced in this lesson as a problem-solving process.

Wind and Wind Energy

Wind is caused by the unequal heating of Earth's surface due to Earth's rotation and different landforms, such as mountains. The Sun's heat is absorbed by Earth's surface and released back into the atmosphere. Because the surface of the land is uneven and the Sun shines on both land and water, different amounts of heat energy are released back into the air in different areas, causing temperature differentials. Different air temperatures are associated with different pressures. Cool air is denser than warm air, so cool air increases atmospheric pressure and sinks, while warm air reduces pressure and rises; when this air movement occurs, wind is created. Wind speed and direction are affected by atmospheric pressure, with wind blowing from areas of high pressure to low pressure. The National Energy Education Development (NEED) project provides educational resources on wind and wind energy that may be useful to you and your students. See "Wind Energy" at *www.need.org/files/curriculum/infobook/WindP.pdf*.

Wind is a renewable energy source (see definition in Table 4.2, p. 51). Although wind energy supplied only about 5% of the energy in the United States in 2016, according to the U.S. Energy Information Administration (EIA), it is the fastest-growing electricity source in the country. (For more information, see "Installed Wind Capacity" from the U.S. Department of Energy (DOE) at *https://apps2.eere.energy.gov/wind/windexchange/wind_installed_capacity.asp* and "Electricity Generation from Wind" from the U.S. Energy Information Administration (EIA) at *www.eia.gov/kids/energy.cfm?page=wind_home-basics#wind_electricity_generation-basics*).

There is evidence that wind energy has been used for thousands of years to propel boats, pump water, and grind grain. Wind energy was used to power windmills as early as 200 BC in China and the Middle East. The Dutch are known for their work in advancing and refining windmill technologies beginning around 1000 AD when wind power technologies spread to northern Europe. These technologies crossed the ocean with the settlement of the United States and were important in the country's westward expansion because windmills were commonly used to access water. Modern wind turbines use technology similar to that of historic windmills, with blades to capture the wind's kinetic energy. In modern turbines, however, the blades are connected to an electric generator that produces electricity for households and businesses.

Engineering

Students begin to gain an understanding of engineering and other professions related to resource use in this module. In particular, they should understand that engineers are people who design and build products in response to human needs. Engineers apply science and mathematics knowledge to create these designs and solutions. Students should also understand that there are many different types of engineers. For an overview of the various types of engineering professions, see the following websites:

- *www.engineeryourlife.org/?ID=6168*

- *www.nacme.org/types-of-engineering*

- *www.sciencekids.co.nz/sciencefacts/engineering/typesofengineeringjobs.html*

Career Connections

As career connections related to this lesson, you may wish to introduce the following:

- *Urban Planner:* Urban planners work to optimize the effectiveness of a community's land use by developing plans to create communities and plan for growth.

- *Geographer:* Geographers study the Earth's natural land formations and human society, with a focus on the relationship between these phenomena. In particular, they study the characteristics of various parts of the Earth, including physical characteristics and human culture. Many geographers work for the federal government. Teaching and field research are other areas in which geographers work. For more information, see *www.bls.gov/ooh/life-physical-and-social-science/geographers.htm.*

- *Mechanical Engineer:* Mechanical engineers design and build mechanical systems (such as motors) and tools.

- *Electrical Engineer:* Electrical engineers design electrical circuits and computer chips.

- *Civil Engineer:* Civil engineers design bridges, roads, and dams.

- *Computer Engineer:* Computer engineers do work that is similar to that of electrical engineers, but they specialize in computer technology. Much of their work with electrical circuits is on a very small scale, such as in microprocessors.

Engineering Design Process (EDP)

Students should understand that engineers need to work in groups to accomplish their work, and that collaboration is important for designing solutions to problems. In this module, students are challenged to work in teams to complete a variety of tasks and to act as design engineers. They will use the engineering design process (EDP), the same process that professional engineers use in their work. Your students may be familiar with the scientific method but may not have experience with the EDP. Students should understand that the processes are similar but are used in different situations. The scientific method is used to test predictions and explanations about the world. The EDP, on the other hand, is used to create a solution to a problem. In reality, engineers use both processes, and your students' experience will reflect this. A good summary of the similarities and differences between the processes is at *www.sciencebuddies.org/engineering-design-process/engineering-design-compare-scientific-method.shtml*. An additional resource about the EDP is the video "What Is the Engineering Design Process?" at *www.pbslearning media.org/resource/phy03.sci.engin.design.desprocess/what-is-the-design-process*.

A graphic representation of the EDP is provided at the end of this lesson. It may be useful to post this in your classroom. You may want to review each step of the EDP listed on the graphic with students.

COMMON MISCONCEPTIONS

Students will have various types of prior knowledge about the concepts introduced in this lesson. Table 4.3 outlines some common misconceptions students may have concerning these concepts. Because of the breadth of students' experiences, it is not possible to anticipate every misconception that students may bring as they approach this lesson. Incorrect or inaccurate prior understanding of concepts can influence student learning in the future, however, so it is important to be alert to misconceptions such as those presented in the table.

Table 4.3. Common Misconceptions About the Concepts in Lesson 1

Topic	Student Misconception	Explanation
Engineering and the engineering design process (EDP)	Engineers use only the scientific process to solve problems in their work.	The scientific method is used to test predictions and explanations about the world. The EDP, on the other hand, is used to create a solution to a problem. In reality, engineers use both processes. (See Teacher Background Information section on page 52 for a discussion of this topic.)
Weather	The speed and direction of air movements are not measurable.	The direction and speed of air movement can be measured by relatively simple tools such as weather vanes and anemometers.
	Wind is a random movement of air with no discernible cause or predictable pattern.	Wind is caused by differences in atmospheric pressure. When air moves from areas of high to low pressure, winds result. The direction of the wind is influenced by Earth's rotation.
	Landforms do not affect the weather.	Landforms can affect weather patterns by altering wind patterns and evaporation rates, thereby causing changes in temperature, humidity, and precipitation.

PREPARATION FOR LESSON 1

Review the Teacher Background Information, assemble the materials for the lesson, and preview the videos recommended in the Learning Plan Components section below. Have your students set up their STEM Research Notebooks (see pp. 24–26 for discussion and student instruction handout). Students should include all work for the module in the STEM Research Notebook, so you may wish to have them include section dividers in their notebooks. A STEM Research Notebook entry rubric is attached at the end of this lesson for use throughout the module.

For the Scarcity Scramble activity, you need to prepare an envelope for each country with pieces of paper labeled as resources. Each team of students will act as a country, so students should be grouped into six teams for the activity. The following list indicates how many units of each resource are needed for each country. Each slip of paper should

equal 1 unit of the particular resource so that students can trade. For example, the United States envelope should have 9 slips of paper labeled as water, 7 as topsoil, and so on. You will act as the World Bank, lending resources to teams that might run out of resources during the activity. Make slips of paper to create a "bank" of about 30 additional cash units (1 cash unit per slip of paper).

United States

- Water: 9 units
- Topsoil: 7 units
- Energy to produce electricity: 10 units
- Oil/gasoline: 4 units
- Minerals: 2 units
- Housing materials: 7 units
- Human resources: 7 units
- Farm animals: 12 units
- Cash: 15 units

Mali

- Water: 3 units
- Topsoil: 2 units
- Energy to produce electricity: 6 units
- Oil/gasoline: 3 units
- Minerals: 10 units
- Housing materials: 2 units
- Human resources: 3 units
- Farm animals: 2 units
- Cash: 3 units

Chile

- Water: 12 units
- Topsoil: 4 units
- Energy to produce electricity: 8 units
- Oil/gasoline: 2 units
- Minerals: 9 units
- Housing materials: 3 units
- Human resources: 4 units
- Farm animals: 4 units
- Cash: 6 units

Bangladesh

- Water: 6 units
- Topsoil: 10 units
- Energy to produce electricity: 6 units
- Oil/gasoline: 5 units
- Minerals: 3 units
- Housing materials: 8 units
- Human resources: 3 units
- Farm animals: 3 units
- Cash: 4 units

Russia

- Water: 18 units
- Topsoil: 6 units
- Energy to produce electricity: 9 units
- Oil/gasoline: 8 units
- Minerals: 6 units
- Housing materials: 3 units
- Human resources: 7 units
- Farm animals: 7 units
- Cash: 7 units

Canada

- Water: 18 units
- Topsoil: 6 units
- Energy to produce electricity: 10 units
- Oil/gasoline: 5 units
- Minerals: 5 units
- Housing materials: 6 units
- Human resources: 4 units
- Farm animals: 3 units
- Cash: 12 units

The Blown Away Design Challenge requires that you set up a small wading pool (plastic or inflatable) and fill it with a few inches of water for student teams to test and time their boats. An EDP Engineer It! Rubric handout is attached at the end of this lesson for use throughout the module.

LEARNING PLAN COMPONENTS
Introductory Activity/Engagement

Connection to the Challenge: Begin each day of this lesson by directing students' attention to the driving question for the module and challenge: Where could we locate a wind farm that a community would support? Ask students where they think the power for their school and their homes comes from. Create a class list of ideas about energy sources. Next, ask students what other sources of energy they can think of. Create a list of these ideas. Then, introduce the Wind Farm Challenge to students, telling them that they will use what they learn about wind energy and geography to choose a spot for a wind turbine somewhere in the United States. Show students the illustration for the Wind Farm Challenge (attached at the end of Lesson 4 on p. 185).

Social Studies Class: Hold a class discussion about wind energy. Ask students what they know about wind energy, and begin a KWL (Know, Want to Know, Learned) chart with students' ideas and questions about wind energy. Maintain this chart throughout the module, adding student ideas as appropriate. Ask students if they think that a whole town could be powered by wind. Then, tell students that there is a town in the United States that gets all its energy from the wind. Show a video about Rock Port, Missouri (visit YouTube and search for "Innovative Cities Rock Port Missouri" or access the video directly at *www.youtube.com/watch?v=mnifNSrZRUQ*). After watching the video, ask students to share what they learned about wind energy from the video. Begin to fill in the Learned column of the KWL chart.

Science Class: Hold a class discussion about wind, asking students questions such as the following:

- What is the wind?

- What creates wind?

- How do we know the wind is blowing?

- How can we measure how fast the wind blows?

- Is the wind different in different areas of the world? How is it different? Why?

Record student answers on the KWL chart. To further assess students' current understanding of wind energy and other science content in this module, as well as to identify misconceptions that need to be addressed, have students answer the questions on the preassessment attached at the end of this lesson.

Mathematics Connection: Not applicable.

ELA Connection: Show students the video "Moving Windmills: The William Kamkwamba Story," based on the book *The Boy Who Harnessed the Wind* (visit YouTube and search for the video title or access the video directly at *www.youtube.com/watch?v=arD374MFk4w*). After watching the video, have students create an entry in their STEM Research Notebooks in response to the following questions:

- What resources did William use to learn about and design windmills?

- How are the resources available to you the same as the resources William used? How are they different?

- Describe something (such as a product or a process) that you might be interested in designing and building or creating. What resources would you use to figure out how to do this?

Activity/Exploration

Social Studies Class: Students investigate renewable and nonrenewable resources and scarcity in three activities: Enough for Everyone?, Resource Your Day, and Scarcity Scramble.

Enough for Everyone?

Begin this activity by holding a class discussion on resources. Have students brainstorm to create a list of natural resources. Next, introduce the concept of renewable and nonrenewable resources, explaining that renewable resources are those that can be replenished within a relatively short period of time, and nonrenewable resources are those that are depleted with use and cannot be replenished within our lifetime. Ask students to review the list of resources the class made and categorize each as renewable or nonrenewable. Circle the renewable resources in green. Ask students if they can think of any other renewable or nonrenewable resources to add to the list.

In this activity, students experience resource scarcity and the concept of renewable and nonrenewable resources. Divide students into teams of four. Give each team a "resource pot" of candy. Tell students that this candy represents the resources they use each day. On each team, assign students numbers from 1 to 4. Designate one color of candy to represent renewable resources; the other colors represent nonrenewable resources. Tell students that, starting with student 1, they will each draw a number from the bag with slips of paper, and then draw that number of candies from the candy bag. After student 1 draws, the student should count the number of renewable resources and the number of nonrenewable resources he or she chose and record that number on the data sheet. Then, the student should return the pieces representing renewable resources to the candy bag but place the nonrenewable pieces in the discard bag (the plastic bag). Next, each student in turn from 2 to 4 draws and repeats this procedure. Continue with additional rounds until a team member finds that insufficient candy remains in the bag to meet his or her resource needs based on the number drawn. At that point, the game ends for that team.

Discuss with the class how fast each team ran out of nonrenewable resources. How many rounds of drawing did it take for each team? Ask them to consider whether there are resources that we use every day that could run out at some point (e.g., fossil fuels such as oil and natural gas). Ask students why there were fewer renewable than nonrenewable resources in the bag (e.g., our cars and power plants aren't designed to use renewable resources; they are difficult to harness and transport). Introduce the concept of scarcity to students. Have students share their ideas about resources that might be scarce for some people.

Resource Your Day

Students continue to work in groups of four in this activity. Using the Resource Your Day handout, students identify several resources they use and encounter at points throughout their day. Then, they identify the source of this resource, conducting internet research if necessary, and classify that source as renewable or nonrenewable.

Scarcity Scramble

Group students into six teams. The goal is for each team to end up with sufficient resources for their country (see the Scarcity Scramble student handouts attached at the end of this lesson). Give each team an envelope representing a country. You may wish to have students create country flags as identifiers for their countries' groups. Each envelope contains slips of paper representing some combination of resources and cash. Students need to barter with other countries (teams) to try to collect all the items on the sufficient resources list. You will act as the World Bank for nations that run out of money to purchase resources and are no longer able to trade. Students may "borrow" up to 5 cash units from you. Give students 15 minutes to complete the activity, and then hold a class discussion, giving each group an opportunity to share its results.

Science Class: Students investigate the role of each of Earth's systems in creating wind in the Earth's Spheres activity. Then, they use the EDP to solve a problem in the Blown Away activity.

Introduce the activities by asking students how wind is used. Create a list of student ideas. Next, ask students where the wind comes from. Students may be surprised that the wind actually originates with the Sun (see Teacher Background Information section on p. 52). Show the video "Bill Nye the Science Guy on Wind," which explains wind (visit YouTube and search for this title or access the video directly at *www.youtube.com/watch?v=uBqohRu2RRk*).

Earth's Spheres

Introduce this activity by asking students to identify what they learned from the video about what contributes to wind (e.g., the Sun, the land, the air). Emphasize to students that different parts of Earth and its surroundings interact to create wind. Introduce the idea that Earth is composed of different systems, and ask students to share their ideas about what the word *system* means. Create a class list of student ideas. Guide students to an understanding that a system is a set of parts that interact with one another to create a complex whole. Ask students to share where they have used the term *system* before (e.g., digestive system, computer system, solar system). Direct students' attention to body systems (e.g., digestive, respiratory, circulatory, and nervous systems), and ask students to describe how each meets the definition of a system. Next, ask students whether any of these systems can function alone. Introduce the idea that systems often interact with one another and that Earth, like the human body, is made of systems that work together in complex ways.

Show students *The Blue Marble*, a photo of Earth that was taken from the *Apollo 17* spacecraft (this photo can be found at *http://earthobservatory.nasa.gov/IOTD/view.php?id=1133*). Ask students to identify the parts of Earth they see in this picture (e.g., land, water, clouds). Create a class list of student responses. Tell students that they are observing some of Earth's systems in this photo. Ask students if there are any other parts of Earth that are not as easy to observe in the picture. Tell students that there are four major parts of Earth that work together: land, water, air, and living things. Introduce the concepts of Earth's spheres to students by telling them that we call each of these major parts or systems *spheres*.

Ask students to share ideas about what a sphere is. Guide students to understand that a sphere is a round ball, like our Earth, and that Earth's spheres can be thought of as spheres inside one another; for example, the atmosphere can be thought of as a sphere, with the other Earth spheres inside it. Emphasize to students that these spheres are Earth systems, and although they are not all shaped like a ball, the idea of spheres is a model that we use to understand Earth's systems. Use the Earth's Spheres image (p. 76) to introduce the following four spheres:

- Atmosphere: the layer of gases surrounding Earth, commonly referred to as air

- Hydrosphere: all the water on Earth's surface, including the oceans, lakes, rivers, groundwater, glaciers, and polar ice caps

- Biosphere: the outer parts of Earth that are occupied by living things, including plants, animals, insects, and microscopic organisms such as bacteria; the biosphere includes parts of the surface, atmosphere, and hydrosphere

- Lithosphere: Earth's solid outer surface, consisting of its crust and mantle and including the continents, the floor of the ocean, rocks, metal, and dust

Show students the Earth's Spheres handouts labeled as Pictures 1–5 (pp. 77–81), and ask them to identify the spheres they observe in each picture. Next, ask students to look out the classroom window and share ideas about how they see parts of the four spheres interact with each other. Create a class list of interactions between things that are part of different spheres. For example, rain is absorbed into the ground (lithosphere) and provides water to plants (biosphere); plants (biosphere) release oxygen and water vapor into the air (atmosphere); humans (biosphere) expend energy they get from plants and other animals (biosphere) to build buildings), parking lots, and roads (lithosphere); wind (atmosphere) causes tree branches (biosphere) to fall; humans (biosphere) release gases into the air (atmosphere) when they breathe out.

Have students create a STEM Research Notebook entry providing the definitions for the four different spheres along with a simple picture for each (e.g., wavy lines to depict wind for atmosphere, waves for hydrosphere, a flower and an insect for biosphere, and a mountain for lithosphere).

Next, post copies of Pictures 1–5 at five stations around the room. Divide students into five groups to rotate among the picture stations, and distribute five copies of the Connect the Spheres handout (p. 82) per student, so that each will have one handout per image.

Allow students to take their STEM Research Notebooks with them so they can refer to the sphere definitions if needed. Give students 5 minutes at each station to identify interactions in the pictures by drawing arrows and writing labels on the Connect the Spheres handouts. After students have visited all picture stations, have them share with the class some of the interactions they observed for each picture.

Tell students that they created models of interactions using their arrows and explanations. Ask students to share ideas about what a model is and provide examples of models, creating a class list of examples. Next, have students share ideas about why we make models. Develop a class definition of model based on student responses. Guide students to understand that models are useful ways to "see" and understand complex things and make predictions about them, but that models have limitations. Have students share ideas about how models are limited (e.g., they cannot include every detail).

Now, tell students that the class will work together to create a model of how Earth's spheres interact and how wind is a part of that. Ask students to recall what they learned about the source of wind from the video they watched earlier, revisiting the KWL chart. Ask students to name one thing that is essential for wind to blow. Guide students to understand that the source of the wind can be traced back to the Sun and the way it heats Earth. It is important for students to understand that some parts of Earth get hotter than others and that these different temperatures are released back into the atmosphere, creating different air temperatures in different areas. You may wish to use an area where land meets water as an example. Since land usually absorbs and releases heat faster than water, the air over the land is warmer than the air over the water. The warmer air rises more quickly, and the cooler air rushes in to take its place, creating wind.

Have students work in teams to investigate Earth's spheres and create diagrams that will serve as models of the spheres. Group students in teams of three or four, and assign each team to investigate one sphere (you will probably need to assign the same sphere to more than one team). Each student team should investigate its assigned sphere and create labeled diagrams on chart paper depicting the sphere and its interaction with wind. These models should show both how the team's assigned sphere contributes to the formation of wind and how that sphere is affected by the wind. Students should use the Earth's Spheres Graphic Organizer to record information about their spheres and plan their drawings.

Have each team present its assigned sphere and model. After the team presentations, hold a class discussion about how the diagrams showed the interaction of the Earth's spheres with one another. Add to the KWL chart on wind. Discuss the diagrams as models and have students offer ideas about the limitations of these models.

Blown Away Design Challenge

In this activity, students use the EDP to create a solution to a challenge to use wind to move boats (see Teacher Background Information section on p. 54 for more information on the EDP). Show students the EDP graphic attached at the end of this lesson as you discuss the process with the class. It may be useful to post this graphic in your classroom as a reminder while they work on the challenge. In this challenge, students will work in teams to create small sailboats and sail them across a wading pool with the wind from an electric fan.

Introduce the challenge by asking the class to brainstorm about how wind has been used through history (e.g., windmills, sailboats). Tell students that they are going to be challenged to create a type of sailboat today, but they will have to come up with their own designs. Ask students to share ideas about what they need to consider in building a sailboat (e.g., it must be watertight, the sails must catch the wind). Tell students that they will have a set of materials that they can use however they like in the challenge. Review the rules and EDP Engineer It! handouts before students begin work. Announce the time limit you have decided on for the challenge (each team must have time to test its design once and then redesign after testing).

After the designs are finished, have students hold a race or time trials. Give each team an opportunity to present its design and explain its design choices to the class. As a class, experiment with different wind intensities and directions, by changing the position and setting of the fan, to gauge the effect of the wind changes on the boats' motion. Hold a class discussion about why these wind changes might occur.

Mathematics Connection: Using the fictional resource allocations in the Scarcity Scramble activity, have students use addition, subtraction, multiplication, and division to determine how the total resources available could be divided equally among groups. Next, have students research the population of each country. Ask students to consider whether dividing resources equally among countries means that each person across the globe would have access to the same resources. As a class, develop an equation for dividing resources equally among people and, as an optional extension, among countries based on population size.

ELA Connection: Have students create a STEM Research Notebook entry in which they reflect on the Scarcity Scramble. You may wish to prompt students with questions such as the following:

- Do you think the way resources were divided is fair? Why or why not?

- Why do you think some countries have so much less than others?

- Do you think wealthier countries should help countries that don't have adequate resources?

A variety of fiction and nonfiction texts relevant to the lesson topics are available and may be explored throughout the module. You may wish to have students create notebook entries to reflect on the readings and how they may connect to the Wind Farm Challenge. Following are some suggested literature connections:

- *The Boy Who Harnessed the Wind*, Young Readers Edition by William Kamkwamba and Bryan Mealer (Dial Books for Young Readers; ISBN: 978-0803740808).

- *Night of the Twisters*, by Ivy Ruckman (HarperCollins; ISBN: 978-0064401760)

- *Earth's Wild Winds*, by Sandra Friend (21st Century; ISBN: 978-0761326731)

- *Wind and Air Pressure: Measuring the Weather*, by Alan Rodgers (Heinemann; ISBN: 978-1432900816)

- *Wind and the Earth*, by Nikki Bundey (Carolrhoda Books; ISBN: 978-1575054704)

- *Generating Wind Power*, by Niki Walker (Crabtree; ISBN: 978-0778729273)

- *Wind Power*, by Tea Benduhn (Gareth Stevens; ISBN: 978-0836893649)

- *Energy Island: How One Community Harnessed the Wind and Changed Their World*, by Allan Drummond (Farrar, Straus and Giroux; ISBN: 978-0374321840)

- *Poetry for Young People: Robert Frost*, by Gary D. Schmidt and Henri Sorenson (Sterling; ISBN: 978-1402754753)

Explanation

Social Studies Class: Students need to understand the difference between renewable and non-renewable resources and the concept of scarcity for this module. While the activities in this lesson are designed to illustrate these concepts, students should be able to articulate definitions and identify various resources as being either renewable or non-renewable. You may wish to discuss the role of recycling with students in terms of its effect on resource availability.

Since students will conduct internet research to complete activities in this module, provide them with some basic guidelines on conducting internet searches for information. First, direct them to use a safe and reliable search engine (if your school uses a particular kid-friendly search engine, direct students there). Then, encourage them to use the following steps:

1. Enter one or more key words in the search bar, and search for these terms.

2. Skim the list of sites that appears.

3. Preview sites that you think may be helpful by going to the site and quickly reviewing the information there.

4. Spend your time on sites that your preview shows may be helpful.

5. Focus on credible sites (e.g., government websites or educational websites from known sources) instead of blogs and sources that you don't recognize.

6. Try using different search terms if your first search didn't lead you to the information you want.

Science Class: Students should have developed a conceptual understanding of how wind works from the activities they have done, such as Earth's Spheres. They should understand that wind energy actually relies on the Sun and that not all areas have the same amount of wind. Areas near water and flat areas, for instance, tend to be windier. Students need to understand that Earth's systems interact in complex ways to create weather patterns including wind. You may find the following literature connections useful as you teach students about wind and landforms:

- *Wind and Air Pressure: Measuring the Weather,* by Alan Rodgers (Heinemann; ISBN: 978-1432900816)

- *Wind and the Earth,* by Nikki Bundey (Carolrhoda Books; ISBN: 978-1575054704)

As a class, discuss elements of collaboration and teamwork and have students share their ideas about what they can do to be good collaborators. Create a list of class guidelines for being a good teammate, and post it on chart paper in the classroom for use throughout the module. A collaboration rubric for the Blown Away Design Challenge is provided at the end of this lesson.

STEM Research Notebook Prompt

Have students create a STEM Research Notebook entry in which they reflect on how different wind intensities and directions affected the sailboats' movement in the Blown Away activity and record their ideas about why wind speed and direction change.

Mathematics Connection: Review division and interpreting remainders with students in support of their calculations regarding resource allocation (see Activity/Exploration section on p. 58).

ELA Connection: By fifth grade, students are exploring increasing numbers of nonfiction texts in a variety of disciplines and content areas. Students may experience the concepts in this lesson through reading a variety of nonfiction and fiction texts, although most of the suggested literature connections provided for this lesson are nonfiction. This

provides an opportunity for students to investigate the characteristics of fiction versus nonfiction texts. You may wish to emphasize the points in Table 4.4 as you introduce the literature connections for the module. Students may add more items as they brainstorm.

Table 4.4. Characteristics of Nonfiction and Fiction Texts

Nonfiction	Fiction
Conveys facts and information	Tells a story
Is read to learn	Is read to enjoy
May include photos, charts, and graphs	May include illustrations
Presents information and directions	Has characters, a setting, and a plot
May have a table of contents, glossary, and index	May have separate chapters

Elaboration/Application of Knowledge

Social Studies Class: Have students conduct a renewable resource scavenger hunt in their homes to create a list of where renewable resources are harnessed (e.g., windows that face the Sun to capture warmth in the winter, solar-powered calculators or watches).

STEM Research Notebook Prompt

Students should respond to the following prompt in their Research Notebooks: *In this lesson, you experienced the way resources are distributed unequally throughout the world. Think about resources in your town or city. How are they distributed among people? Is it equal? How do people access resources such as water, food, and recreational areas in various neighborhoods? Are there differences?*

Science Class: Ask students to share their ideas about how scientists learn about Earth's spheres and how they interact. Emphasize to students that scientists use observations and measurements over time to learn about Earth's spheres. Create a chart with columns for each sphere—atmosphere, lithosphere, hydrosphere, biosphere—and have students offer ideas about how scientists observe and measure features of each sphere. Next, have students share ideas about what instruments scientists could use to make their observations and measurements. An option is to take students on a walk around the school grounds or neighborhood to identify and observe spheres, and suggest ways that they could learn more about them over time using measurements. Students should track their observations and ideas in their STEM Research Notebooks.

Mathematics Connection: Have students look for examples of resource distribution data in the news (e.g., the gross domestic product (GDP) of various countries, data for mineral-exporting countries).

ELA Connection: Have students create a cover and table of contents for their STEM Research Notebooks.

Evaluation/Assessment

Students may be assessed on the following performance tasks and other measures listed.

Performance Tasks

- Enough for Everyone? handout

- Resource Your Day handout

- Scarcity Scramble handouts

- Blown Away Engineer It! handouts

Other Measures

- STEM Research Notebook Entry Rubric

- Engineer It! EDP Rubric

- Earth's Spheres Rubric

- Blown Away Design Challenge Collaboration Rubric (optional)

- Participation in class discussions

INTERNET RESOURCES

"Wind Energy"
- *www.need.org/files/curriculum/infobook/WindP.pdf*

"Installed Wind Capacity"
- *https://apps2.eere.energy.gov/wind/windexchange/wind_installed_capacity.asp*

"Electricity Generation from Wind"
- *www.eia.gov/kids/energy.cfm?page=wind_home-basics#wind_electricity_generation-basics*

Overview of the various types of engineering professions
- *www.engineeryourlife.org/?ID=6168*

- *www.nacme.org/types-of-engineering*
- *www.sciencekids.co.nz/sciencefacts/engineering/typesofengineeringjobs.html*

Information on geography career
- *www.bls.gov/ooh/life-physical-and-social-science/geographers.htm*

Summary of differences between EDP and scientific process
- *www.sciencebuddies.org/engineering-design-process/engineering-design-compare-scientific-method.shtml*

"What Is the Engineering Design Process?" video
- *www.pbslearningmedia.org/resource/phy03.sci.engin.design.desprocess/what-is-the-design-process/*

"Innovative Cities Rock Port Missouri" video
- *www.youtube.com/watch?v=mnifNSrZRUQ*

"Moving Windmills: The William Kamkwamba Story" video
- *www.youtube.com/watch?v=arD374MFk4w*

"Bill Nye the Science Guy on Wind" video
- *www.youtube.com/watch?v=uBqohRu2RRk*

The Blue Marble photo of Earth from *Apollo 17*
- *http://earthobservatory.nasa.gov/IOTD/view.php?id=1133*

Name: _____

STUDENT HANDOUT

SCIENCE PREASSESSMENT

1. Which of the following statements about wind is true?

   a. Wind is caused only by the movement of Earth.
   b. Since wind is air, it cannot be measured.
   c. Wind is caused by the interaction of warm air and cold air.
   d. Wind energy is a kind of fossil fuel.

2. Which of the following is not an Earth system?

   a. biosphere
   b. megasphere
   c. lithosphere
   d. atmosphere

3. Scientists and engineers use a process called the _____ to learn about the natural world.

   a. science quest
   b. experiment process
   c. scientific method
   d. hypothesis system

4. The following is an example of a hypothesis:

   a. Trees have leaves.
   b. Sunflowers grow in the sun.
   c. If I brush my teeth every day, I will not get cavities.
   d. Brushing my teeth every day is good for my teeth.

5. Identify the variables in the following question: *If I chew a piece of gum longer, can I blow bigger bubbles?*

   a. The brand of gum and the size of the bubbles
   b. The amount of time you chew the gum and the size of bubbles
   c. The color of the gum and the brand of the gum
   d. The size of your teeth and the size of the bubbles

6. Wind energy is a _____:

   a. renewable energy source
   b. fossil fuel
   c. nonrenewable energy source
   d. kind of energy that we do not use in the United States

Wind Energy, Grade 5

Name: _____

STUDENT HANDOUT, PAGE 1

ENOUGH FOR EVERYONE?

INSTRUCTIONS

1. Your teacher will assign each student in your group a number from 1 to 4. This indicates who should go first, second, third, and fourth.

2. Your teacher will tell you what color represents a renewable resource. The other colors represent nonrenewable resources.

3. When your teacher signals your team to begin, the first student should draw a number from the number bag. This number is your resource need. Read this number and, without looking, draw that number of candies from the candy bag.

4. Count the number of renewable resources, and record that in the table below. Return the renewable resources to the bag.

5. Next, count the number of nonrenewable resources, and record that in the table below. Place the nonrenewable resources in the discard bag (the plastic bag).

6. Have the second, third, and fourth students on your team repeat these steps.

7. Continue to take turns until you can no longer meet your resource needs. When this happens, you are out of the game. Once all students are unable to meet their resource needs, the game is over.

Round #	Number of Renewable Resources	Number of Nonrenewable Resources	Total Number of Resources Drawn
1			
2			
3			
4			
5			
6			
7			

ENOUGH FOR EVERYONE?

Answer the following questions:

1. In what round were you unable to meet your resource needs?

   _____

   _____

2. How might this have been different if more renewable resources were available?

   _____

   _____

   _____

3. What kind of resources that we use daily might be nonrenewable like the candies you ran out of?

   _____

   _____

   _____

   _____

   _____

4. Why do you think there were more nonrenewable resources in the bag than renewable resources?

   _____

   _____

   _____

   _____

   _____

STUDENT HANDOUT

RESOURCE YOUR DAY

In this activity, you will name resources associated with each of the activities listed below. You will need to identify the source of that resource and determine whether it is a renewable or nonrenewable resource. Circle the renewable resources and underline the nonrenewable resources.

Good morning! It's a school day. The first task of your day is to get to school. Name three resources you encounter or use in getting to school:

1. Resource: Where this is used: Source:

2. Resource: Where this is used: Source:

3. Resource: Where this is used: Source:

You've gotten to school; now you are in social studies class. Name three resources you encounter or use in your classroom:

1. Resource: Where this is used: Source:

2. Resource: Where this is used: Source:

3. Resource: Where this is used: Source:

It's lunch time! Name three resources you encounter or use during lunch:

1. Resource: Where this is used: Source:

2. Resource: Where this is used: Source:

3. Resource: Where this is used: Source:

You're home again. You have some social studies homework, and you want to spend some time relaxing with your family. Name three resources you encounter or use during your evening at home:

1. Resource: Where this is used: Source:

2. Resource: Where this is used: Source:

3. Resource: Where this is used: Source:

Name: _____   Country Name: _____

STUDENT HANDOUT, PAGE 1

SCARCITY SCRAMBLE

Your team's goal is to collect sufficient resources for your country. Each team begins with different items in its envelope. You may trade items with other teams or use your cash to purchase items you are missing. If you run out of resources to trade or cash, you may borrow up to 5 cash units from the World Bank (your teacher).

The following are the resources you will need to successfully complete the challenge; they represent sufficient resources for the people in your country:

- Water: 10 units
- Topsoil: 5 units
- Energy to produce electricity: 10 units

- Oil/gasoline – 8 units
- Minerals: 5 units
- Housing materials: 4 units

- Human resources: 5 units
- Farm animals: 5 units
- Cash: 10 units

PREACTIVITY ACCOUNTING

You have a combination of resources in your bag and some cash. Record in the table below what is in your bag and what you need.

Resource	Amount You Have	Amount You Need (how much you need to get from another team to have a complete set of resources)	Amount Extra You Have (how much you can use to trade or buy other resources)

Name: _____ Country Name: _____

STUDENT HANDOUT, PAGE 2

SCARCITY SCRAMBLE

POSTACTIVITY ACCOUNTING

When your teacher tells you to stop, fill in the table below to reflect what your team has now.

Resource	Target Amount	Amount You Have	Difference (use + to indicate more than needed and – to indicate less than needed)
Water	10 units		
Topsoil	5 units		
Energy to produce electricity	10 units		
Oil/gasoline	8 units		
Minerals	5 units		
Housing materials	4 units		
Human resources	5 units		
Farm animals	5 units		
Cash	10 units		
Amount borrowed from the World Bank			

Name: _____   Country Name: _____

STUDENT HANDOUT, PAGE 3

SCARCITY SCRAMBLE

Answer the following questions:

1. Did all the countries have equal access to resources? Why or why not?

   _____

   _____

2. What happened to countries with the least amount of resources?

   _____

   _____

3. What are some reasons that some countries had difficulty filling their resource needs?

   _____

   _____

4. What countries did not have difficulty filling their resource needs?

   _____

   _____

EARTH'S SPHERES

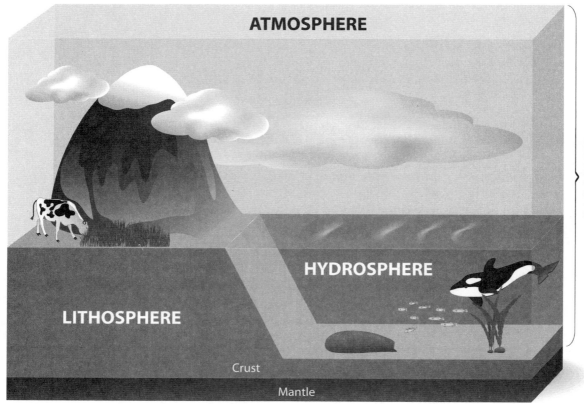

Note: A full-color version of this image is available on the book's Extras page at *www.nsta.org/roadmap-wind.*

EARTH'S SPHERES: PICTURE 1

Note: A full-color version of this image is available on the book's Extras page at *www.nsta.org/roadmap-wind.*

EARTH'S SPHERES: PICTURE 2

Note: A full-color version of this image is available on the book's Extras page at *www.nsta.org/roadmap-wind*.

EARTH'S SPHERES: PICTURE 3

Note: A full-color version of this image is available on the book's Extras page at *www.nsta.org/roadmap-wind.*

EARTH'S SPHERES: PICTURE 4

Note: A full-color version of this image is available on the book's Extras page at *www.nsta.org/roadmap-wind.*

EARTH'S SPHERES: PICTURE 5

Note: A full-color version of this image is available on the book's Extras page at *www.nsta.org/roadmap-wind*.

STUDENT HANDOUT

CONNECT THE SPHERES

Picture #: _____

Draw arrows to show where you see Earth's spheres interacting in the picture. Label the arrows to explain what the interaction is.

Atmosphere   **Hydrosphere**

Biosphere   **Lithosphere**

Name: _____   Assigned Sphere: _____

STUDENT HANDOUT

EARTH'S SPHERES GRAPHIC ORGANIZER

Complete this graphic organizer before you create your team's diagram.

What does the prefix for the name of this sphere mean?
(circle one: atmo-, hydro-, litho-, or bio-):

_____

This sphere is made up of the following:

Three examples of items in my team's assigned sphere:

This sphere helps create the wind by:

This sphere is changed by wind in these ways (give two examples)

ENGINEERING DESIGN PROCESS

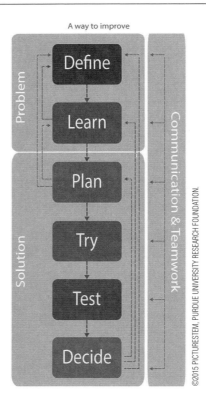

Name: _____

BLOWN AWAY DESIGN CHALLENGE

Imagine that you and your team are part of a reality television show in which you are stranded on an island. You must build a boat to travel to a neighboring island as quickly as possible. Your challenge is to build a boat that can be powered by wind using the given supplies. You have a limited amount of time to complete and test your design. (Your teacher will tell you how much time you have.) Your design must meet the following requirements:

1. You may use only the materials supplied. (You do not need to use all of them.) The paper clips may not be used in the construction of your boat; these will be the load your boat will carry.

2. Each member of your team must track the team's design process using his or her own Engineer It! handouts.

3. Your boat must be powered only by wind energy (supplied by the fan).

4. Your boat must be able to carry a load of eight paper clips representing your team members to the other island (the other side of the wading pool).

5. The paper clip load must not blow off your boat.

6. When your team is ready to test its design, it will have 1 minute to test using the wind source (the fan).

7. Your boat design will be judged by how quickly it reaches the other side and how safely it transports your passengers (paper clips). In the case of a tie between teams' designs, more paper clips will be added to each boat and the trial repeated.

8. Be prepared to tell the class why you designed your boat the way you did.

Wind Energy Lesson Plans

Name: _____ Team Name: _____

STUDENT HANDOUT, PAGE 1

ENGINEER IT! BLOWN AWAY

Note: The "try" phase is assessed using the collaboration rubric; the "share" phase is assessed using the presentation rubrics.

Step 1: (Define) State the problem (what are you trying to do?):

Step 2: (Learn) What solutions can you and your team imagine?

Step 3: (Plan) Make a sketch of your design here and label materials:

Step 4: (Try) Build it! Did everyone on your team have a chance to help?

NATIONAL SCIENCE TEACHERS ASSOCIATION

Name: _____   Team Name:_____

STUDENT HANDOUT, PAGE 2

ENGINEER IT! BLOWN AWAY

Step 5: (Test)   Test your design. How did it work? Could it work better?

Step 6: (Decide)   This is your chance to use your test results to decide what changes to make. What would you change to make your boat work better?

Step 7: Share your design! What features of your design do you want to point out to the class?

STEM Research Notebook Entry Rubric (10 points possible)

Name: _____   Date: _____

Below Standard (1–2)	Approaching Standard (3–5)	Meets Standard (6–8)	Exceeds Standard (9–10)
• The written work shows misconceptions and a lack of understanding.	• The written work shows a limited understanding of concepts.	• The writing follows the basic requirements, shows understanding of concepts, but does not go beyond.	• The writing goes beyond the basic requirements and shows in-depth understanding of concepts.
• Little reflection is in evidence.	• Limited reflection overall.	• The work shows in-depth reflection.	• The work shows in-depth reflection throughout the learning process.
• The entry is unfinished.	• The notebook has some of the components expected, including dates and labels.	• The entry has most of the components expected, including dates and labels.	• Notebook has all the components expected, including dates and labels.
• Dates and labels are missing.	• Work is somewhat organized but could be neater.	• Work is neat and organized.	• Work is neat and very well organized.
• Work is disorganized or not neat.	• The entry uses little color and few diagrams where these would be appropriate.	• The entry uses color and labeled diagrams where appropriate.	• The use of color and labeled diagrams enhances understanding.
• No color or diagrams used where these would be appropriate.			

SCORE: _____

Engineer It! EDP Rubric (20 points possible)

Name: _____

EDP Step	Below Standard (0–1)	Approaching Standard (2–3)	Meets or Exceeds Standard (4–5)	Student Score
DEFINE	• The problem is unclear or is not stated.	• The problem is stated but may be unclear or may have grammatical errors.	• The problem is stated clearly and using correct grammar.	
LEARN	• Little or no evidence that student participated in team brainstorming or research. Few or no ideas recorded.	• Some evidence of student participation in team brainstorming or research. Ideas may be expressed incompletely or be unclear.	• Evidence that student actively participated in team brainstorming or research. Several ideas are expressed clearly and completely.	
PLAN	• Student sketch is missing, is not understandable, or lacks most design elements and labels.	• Student provided a sketch that has some of the design elements and labels.	• Student provided a sketch that includes all design elements, is easy to understand, and is clearly labeled.	
TEST/ DECIDE	• Little or no evidence that student participated in testing the design or student had no ideas about how to improve the design.	• Evidence that student participated in testing the design and provided some ideas for improving the design; however, ideas may be incomplete or difficult to understand.	• Evidence that student participated in testing the design and provided clear, understandable, and innovative ideas for improving the design.	

TOTAL SCORE: _____

Team Name: _____

Earth's Spheres Rubric (15 points possible)

	Below Standard (0–1)	Approaching Standard (2–3)	Meets or Exceeds Standard (4–5)	Team Score
CONTENT	• Information is not complete; does not include all required information.	• Most information is provided, but some information is missing or inaccurate	• All information is provided and is accurate	
PLANNING	• Content is not labeled, is not logically organized and is difficult to understand. Evidence that not all team members participated in preparation.	• Content is labeled and is somewhat organized but may be difficult to understand in some places. Evidence that most, but not all, team members participated in preparation.	• Content is labeled, in logical order, and easy to understand. Evidence that all team members participated in preparation.	
PRESENTATION	• Information is unclear, difficult to understand, and incomplete. Only one team member participated in the presentation. Team members were unable to respond to questions appropriately.	• Information is complete, but may be difficult to understand at times. Most team members, but not all, participated. Team was able to respond to some but not all questions appropriately.	• Information is complete and easily understandable. All team members participated. Team members responded to all questions appropriately.	

TOTAL SCORE: _____

Blown Away Design Challenge Collaboration Rubric (15 points possible)

Student Name: _____   Team Name: _____

Individual Performance	Below Standard (0–1)	Approaching Standard (2–3)	Meets or Exceeds Standard (4–5)	Student Score
INDIVIDUAL ACCOUNTABILITY	• Student is unprepared. • Student does not communicate with team members and does not manage tasks as agreed on by the team. • Student does not complete or participate in project tasks. • Student does not complete tasks on time. • Student does not use feedback from others to improve work.	• Student is usually prepared. • Student sometimes communicates with team members and manages tasks as agreed on by the team, but not consistently. • Student completes or participates in some project tasks but needs to be reminded. • Student completes most tasks on time. • Student sometimes uses feedback from others to improve work.	• Student is consistently prepared. • Student consistently communicates with team members and manage tasks as agreed on by the team. Student discusses and reflects on ideas with the team. • Student completes or participates in project tasks without being reminded. • Student completes tasks on time. • Student uses feedback from others to improve work.	
TEAM PARTICIPATION	• Student does not help the team solve problems; may interfere with teamwork. • Student does not express ideas clearly, pose relevant questions, or participate in group discussions. • Student does not give useful feedback to other team members. • Student does not volunteer to help others when needed.	• Student cooperates with the team but may not actively help solve problems. • Student sometimes expresses ideas, poses relevant questions, elaborates in response to questions, and participates in group discussions. • Student provides some feedback to team members. • Student sometimes volunteers to help others.	• Student helps the team solve problems and manage conflicts. • Student makes discussions effective by clearly expressing ideas, posing questions, and responding thoughtfully to team members' questions and perspectives. • Student gives useful feedback to others so they can improve their work. • Student volunteers to help others if needed.	
PROFESSIONALISM AND RESPECT FOR TEAM MEMBERS	• Student is impolite or disrespectful to other team members. • Student does not acknowledge or respect others' ideas and perspectives.	• Student is usually polite and respectful to other team members. • Student usually acknowledges and respects others' ideas and perspectives.	• Student is consistently polite and respectful to other team members. • Student consistently acknowledges and respects others' ideas and perspectives.	

TOTAL SCORE: _____

Lesson Plan 2: Where's the Wind?

In this lesson, students make connections between the geography of the United States and the availability of wind resources. They explore the scientific process and measurement of natural phenomena. By investigating a variety of maps, including topographic, relief, offshore, and road maps, students gain a conceptual understanding of the geographic diversity of the United States and how that diversity is expressed in maps. Students compare and contrast the availability of wind resources across the United States and investigate the locations of current wind farms. They also gain a broad understanding of the role wind turbines play in electricity production. Students consider various tools scientists use to measure wind and then use the EDP and the scientific process to design and test an anemometer.

ESSENTIAL QUESTIONS

- How do Earth's landforms affect the availability of wind that can be harnessed for wind farms?

- How do wind resources vary across the continental United States?

- How can we use the scientific process as part of the EDP?

- How is wind measured?

ESTABLISHED GOALS AND OBJECTIVES

At the conclusion of this lesson, students will be able to do the following:

- Identify various types of maps, including topographic, relief, offshore, and road maps

- Apply their understanding of map symbols and conventions to find landforms and geographic locations on maps

- Apply their understanding of the differences in geography across the United States to make predictions about wind resource availability

- Identify geographic features of a U.S. region and create a map of these features

- Apply their map-reading skills to locate and identify existing wind farms in a region of the United States

- Identify tools scientists use to measure natural phenomena in each of Earth's spheres

- Identify and describe the steps of the scientific method

- Develop a testable question and a hypothesis for a scientific investigation

- Create a plan for a scientific investigation using the scientific method

- Describe how an anemometer differs from a weather vane and what purpose each plays in wind measurement

- Apply their understanding of anemometer design and function and of the EDP to design and build a functioning anemometer

- Collaborate with peers to create a solution to a problem

TIME REQUIRED

- 7 days (approximately 45 minutes each day; see Tables 3.7–3.8, pp. 42–43)

MATERIALS

Required Materials for Lesson 2

- STEM Research Notebooks

- Internet access for student research and viewing videos

- Handouts (attached at the end of this lesson)

- Topographic maps for your area and for the five regions of the United States (downloadable maps are available at *www.natgeomaps.com/trail-maps/pdf-quads*)

- Reference map of U.S. regions

- Hand crank flashlight

- Outdoor thermometer (either one that can be mounted outdoors near a window or a remote wireless outdoor thermometer)

- Meter stick

- Scale or balance

- Graduated cylinder or measuring cup

Additional Materials for Marvelous Maps

- U.S. road maps (1 per group)

- Rulers (1 per student)

- U.S. Map Scavenger Hunt handout (1 per student; attached at the end of this lesson)

Additional Materials for Map Me!

- Map Me! handout and blank map (1 per student; attached at the end of this lesson)
- U.S. road map (1 per group)
- Colored pencils (1 set per student)
- Slips of paper with the names of U.S. regions (enough for 1 per team; optional)
- Paper bag (optional)

Additional Materials for How Windy Is the Wind? (per group unless otherwise indicated):

- Box fan (1 per class; optional for use if anemometers cannot be tested outdoors)
- 1 roll of masking tape (per class)
- 8 plastic or paper cups (3 oz. size)
- 8 plastic drinking straws
- 1 pencil with eraser
- 1 pushpin
- 8 small craft sticks
- Play-Doh or clay (about ¼ cup)
- Scissors
- Scotch tape
- 1 piece of cardboard (about 6 × 6 inches)
- Stapler
- Ruler
- How Windy Is the Wind? Engineer It! handouts (1 per student)
- Safety glasses or goggles

SAFETY NOTES

1. All laboratory occupants must wear safety glasses or googles during all phases of this inquiry activity.

2. Immediately wipe up any water on the floor to avoid a slip-and-fall hazard.

3. Keep away from electrical sources when working with water because of the shock hazard.

4. Use caution when operating electrical devices (e.g., fans) because of the potential shock hazard, especially near water.

5. Use caution when working with sharps (scissors, sticks, pins, fan blades, etc.) to avoid cutting or puncturing skin.

6. Make sure all materials are put away after completing the activity.

7. Wash hands with soap and water after completing the activity.

CONTENT STANDARDS AND KEY VOCABULARY

Table 4.5 lists the content standards from the *NGSS, CCSS,* and the Framework for 21st Century Learning that this lesson addresses, and Table 4.6 (p. 99) presents the key vocabulary. Vocabulary terms are provided for both teacher and student use. Teachers may choose to introduce some or all of the terms to students.

Table 4.5. Content Standards Addressed in STEM Road Map Module Lesson 2

NEXT GENERATION SCIENCE STANDARDS

PERFORMANCE EXPECTATIONS

- 3-5-ETS1-1. Define a simple design problem reflecting a need or a want that includes specified criteria for success and constraints on materials, time, or cost.

- 3-5-ETS1-2. Generate and compare multiple possible solutions to a problem based on how well each is likely to meet the criteria and constraints of the problem.

- 3-5-ETS1-3. Plan and carry out fair tests in which variables are controlled and failure points are considered to identify aspects of a model or prototype that can be improved.

DISCIPLINARY CORE IDEAS

ESS3.A. Natural Resources

- Energy and fuels that humans use are derived from natural sources, and their use affects the environment in multiple ways. Some resources are renewable over time, and others are not.

ETS1.A. Defining and Delimiting Engineering Problems

- Possible solutions to a problem are limited by available materials and resources (constraints). The success of a designed solution is determined by considering the desired features of a solution (criteria). Different proposals for solutions can be compared on the basis of how well each one meets the specified criteria for success or how well each takes the constraints into account. (3-5-ETS1-1)

Table 4.5. (*continued*)

CROSSCUTTING CONCEPTS

Scale, Proportion, and Quantity
- Standard units are used to measure and describe physical quantities such as weight, time, temperature, and volume.

Systems and System Models
- A system is a group of related parts that make up a whole and can carry out functions its individual parts cannot.

- A system can be described in terms of its components and their interactions.

Energy and Matter
- Energy can be transferred in various ways and between objects.

Cause and Effect
- Cause and effect relationships are routinely identified, tested, and used to explain change.

Influence of Science, Engineering, and Technology on Society and the Natural World
- Engineers improve existing technologies or develop new ones to increase their benefits, decrease known risks, and meet societal demands.

SCIENCE AND ENGINEERING PRACTICES

Asking Questions and Defining Problems
- Ask questions about what would happen if a variable is changed.

- Identify scientific (testable) and non-scientific (non-testable) questions.

- Ask questions that can be investigated and predict reasonable outcomes based on patterns such as cause and effect relationships.

- Use prior knowledge to describe problems that can be solved.

- Define a simple design problem that can be solved through the development of an object, tool, process, or system and includes several criteria for success and constraints on materials, time, or cost.

Planning and Carrying Out Investigations
- Plan and conduct an investigation collaboratively to produce data to serve as the basis for evidence, using fair tests in which variables are controlled and the number of trials considered.

- Evaluate appropriate methods and/or tools for collecting data.

- Make observations and/or measurements to produce data to serve as the basis for evidence for an explanation of a phenomenon or test a design solution.

- Make predictions about what would happen if a variable changes.

Table 4.5. (*continued*)

> - Test two different models of the same proposed object, tool, or process to determine which better meets criteria for success.
>
> *Analyzing and Interpreting Data*
> - Represent data in tables and/or various graphical displays (bar graphs, pictographs and/or pie charts) to reveal patterns that indicate relationships.
> - Analyze and interpret data to make sense of phenomena, using logical reasoning, mathematics, and/or computation.
> - Compare and contrast data collected by different groups in order to discuss similarities and differences in their findings.
> - Analyze data to refine a problem statement or the design of a proposed object, tool, or process.
> - Use data to evaluate and refine design solutions.
>
> *Using Mathematical and Computational Thinking*
> - Decide if qualitative or quantitative data are best to determine whether a proposed object or tool meets criteria for success.
> - Organize simple data sets to reveal patterns that suggest relationships.
>
> ### COMMON CORE STATE STANDARDS FOR MATHEMATICS
>
> #### MATHEMATICAL PRACTICES
> - MP1. Make sense of problems and persevere in solving them.
> - MP2. Reason abstractly and quantitatively.
> - MP3. Construct viable arguments and critique the reasoning of others.
> - MP4. Model with mathematics.
> - MP5. Use appropriate tools strategically.
>
> #### MATHEMATICAL CONTENT
> - 5.NBT.B.5. Fluently multiply multi-digit whole numbers using the standard algorithm.
> - 5.MD.A.1. Convert among different-sized standard measurement units within a given measurement system (e.g., convert 5 cm to 0.05 m), and use these conversions in solving multi-step, real world problems.
> - 5.G.A.2. Represent real world and mathematical problems by graphing points in the first quadrant of the coordinate plane, and interpret coordinate values of points in the context of the situation.

Table 4.5. (*continued*)

COMMON CORE STATE STANDARDS FOR ENGLISH LANGUAGE ARTS

READING STANDARDS

- RI.5.1. Quote accurately from a text when explaining what the text says explicitly and when drawing inferences from the text.

- RI.5.4. Determine the meaning of general academic and domain-specific words and phrases in a text relevant to a grade 5 topic or subject area.

- RI.5.7. Draw on information from multiple print or digital sources, demonstrating the ability to locate an answer to a question quickly or to solve a problem efficiently.

- RF.5.3. Know and apply grade-level phonics and word analysis skills in decoding words.

- RF.5.4. Read with sufficient accuracy and fluency to support comprehension.

- RF.5.4.a. Read grade-level text.

WRITING STANDARDS

- W.5.2. Write informative/explanatory texts to examine a topic and convey ideas and information clearly.

- W.5.4. Produce clear and coherent writing in which the development and organization are appropriate to task, purpose, and audience.

- W.5.8. Recall relevant information from experiences or gather relevant information from print and digital sources; summarize or paraphrase information in notes and finished work, and provide a list of sources.

SPEAKING AND LISTENING STANDARDS

- SL.5.1. Engage effectively in a range of collaborative discussions (one-on-one, in groups, and teacher-led) with diverse partners on grade 5 topics and texts, building on others' ideas and expressing their own clearly.

- SL.5.1.b. Follow agreed-upon rules for discussions and carry out assigned roles.

- SL.5.1.d. Review the key ideas expressed and draw conclusions in light of information and knowledge gained from the discussions.

- SL.5.6. Adapt speech to a variety of contexts and tasks, using formal English when appropriate to task and situation.

FRAMEWORK FOR 21ST CENTURY LEARNING

Interdisciplinary themes (global awareness; financial, economic, and business literacy; environmental literacy); Learning and Innovation Skills; Information, Media and Technology Skills; Life and Career Skills

---

Table 4.6. Key Vocabulary in Lesson 2

Key Vocabulary	Definition
anemometer	an instrument used for measuring wind speed
contour lines	lines on a map indicating the elevation or height of an area
data	a collection of facts, which can be numbers, measurements, or words
dependent variable	something that changes in an investigation in response to the scientist's actions; the thing that is being measured in an investigation
hypothesis	an explanation that can be tested by investigation
independent variable	the thing that a scientist changes in an investigation to see how it affects something else
map legend	a list of definitions of the symbols used on a map; also called a *key*
map scale	the relationship between a distance on a map and the actual distance
offshore	located in the ocean at a distance from the shore
onshore	on land
relief map	a map that uses shading to show physical characteristics such as hills and valleys
road map	a map showing roads designed to be used for automobile travel
scientific method	a process for conducting research that involves asking a question, forming a hypothesis, conducting an investigation, analyzing data, and drawing conclusions
speed	distance traveled in a certain amount of time; the rate at which something moves
topographic map	a map that uses contour lines to show physical characteristics such as hills and valleys
variable	something that changes in an investigation

TEACHER BACKGROUND INFORMATION

Since students will ultimately be challenged to create a plan for a fictional wind farm, they continue their investigation of wind as an energy source in this lesson by considering how wind turbines operate, placement of wind turbines, the variability of wind around the continental United States, and how wind is measured. Students consider scientific measurement in the context of the scientific method, with a focus on the idea

that scientists and engineers alike ask questions and collect data about those questions. Students apply the scientific method in concert with the EDP to create, test, and improve a model of a wind turbine. You may wish to review the resources on the EDP provided in Lesson 1 (see the Teacher Background Information section on p. 54).

Geography and Map Skills

Students learn about the geographic features of the United States in this lesson. They will become familiar with the five regions in the United States—Northeast, Southwest, West, Southeast, and Midwest—and understand that regions are areas of land grouped according to their location that may have common natural and cultural features. A map of U.S. regions can be found at *http://media.nationalgeographic.org/assets/file/us-regions-map.pdf.*

Students may be familiar with road maps and web mapping services such as Google Maps. Other types of maps include topographic maps, relief maps, political maps, and weather maps. Students explore various types of maps in this lesson to identify landforms in their region.

The U.S. Geological Survey (USGS) provides a number of educational resources for working with maps. These are excellent aids for supporting your students' basic map-reading skills and can be found at *http://education.usgs.gov/primary.html#geoggeneral.*

Wind Turbines

Wind farms are typically developed by companies backed by private investors in the energy industry. Wind energy companies determine the locations of wind farms using meteorological data, information about access to area power lines, and information about environmental and community impact (see Wind Energy Foundation FAQs at *http://windenergyfoundation.org/about-wind-energy/faqs/* for more information). Wind farms are generally located in areas with relatively consistent winds (preferably of about 15 mph) that are unimpeded by geographical formations or by structures. Wind farms are often built in agricultural areas located on plains, in deserts, on the tops of hills, in mountain gaps that funnel wind, or in offshore locations. In 2016, the top wind power producing states in the United States were Texas, Iowa, Oklahoma, Kansas, and California. See the U.S. Energy Information Energy Kids wind energy information at *www.eia.gov/kids/energy.cfm?page=wind_home-basics.*

Anemometers

Anemometers are tools used to measure wind speeds. The most commonly used type of anemometer is a cup anemometer. This type of anemometer is composed of several cups attached to a horizontal arm. The arm is attached to a vertical rod. As the wind blows it moves the cups and rotates the vertical rod. The rotations of the rod are used to calculate wind speed. An image of a cup anemometer is provided in Figure 4.1 (p. 102).

More information about anemometers can be accessed at *www.nationalgeographic.org/encyclopedia/anemometer*.

Career Connections

As career connections related to this lesson, you may wish to introduce the following:

- *Geographer:* Geographers study the Earth's natural land formations and human society, with a focus on the relationship between these phenomena. In particular, they study the characteristics of various parts of the Earth, including physical characteristics and human culture. Many geographers work for the federal government. Teaching and field research are other areas in which geographers work. For more information, see *www.bls.gov/ooh/life-physical-and-social-science/geographers.htm*.

- *Cartographer:* Cartography is a subset of geography. Cartographers interpret geographic information and create maps and charts. Many cartographers have backgrounds in geography and civil engineering. For more information, see *www.bls.gov/ooh/architecture-and-engineering/cartographers-and-photogrammetrists.htm*.

- *Photogrammetrist:* Photogrammetrists use aerial photographs, satellite imagery, and other images to create maps or drawings of geographic areas. This field is closely related to cartography. For more information, see the above website as well as *www.wisegeek.com/what-are-photogrammetrists.htm*.

- *Atmospheric Scientist:* Atmospheric scientists study the weather and climate. This field encompasses several types of careers, including climatologists, meteorologists, and broadcast meteorologists (see specifics on each below). For more information, see *www.bls.gov/ooh/life-physical-and-social-science/atmospheric-scientists-including-meteorologists.htm#tab-2*.

- *Climatologist:* Climatologists study historical weather patterns to identify long-term weather patterns or changes in climate factors and use these data to make predictions, such as anticipated precipitation levels.

- *Meteorologist:* Meteorologists have training in weather patterns and forecasting techniques. They produce weather reports that can be used by the general public or by specific groups, such as farmers or airports.

- *Broadcast meteorologist:* Broadcast meteorologists provide information to the public on current and expected weather conditions for specific locations. They may or may not have training as meteorologists.

Figure 4.1. Cup Anemometer and Wind Vane

Source: National Oceanic and Atmospheric Administration. Public domain. Retrieved from *www.erh.noaa.gov/mhx/tour/ OfficeTourAnemometer.php.*

Note: A full-color version of this image is available on the book's Extras page at *www.nsta.org/roadmap-wind.*

Figure 4.2. Propeller Anemometer

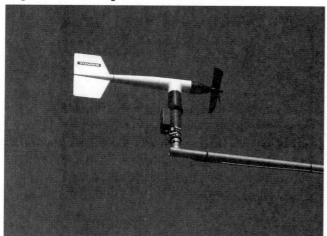

Source: National Oceanic and Atmospheric Administration. Public domain. *www.erh.noaa.gov/mhx/tour/ OfficeTourAnemometer.php.*

Note: A full-color version of this image is available on the book's Extras page at *www.nsta.org/roadmap-wind.*

Wind

Students begin to consider how wind can be measured using an anemometer to measure wind speed. Anemometers typically include a device such as a weather vane that gauges wind direction as well as speed; however, the focus of this lesson is on measuring wind speed using cup anemometers (see Figure 4.1). Another type of anemometer is a propeller anemometer (see Figure 4.2).

You may wish to visit the NOAA education page on weather systems and patterns at *www.noaa.gov/resource-collections/ weather-systems-patterns* for more information about wind, satellite maps of wind, and images of wind patterns in weather events such as a nor'easter.

COMMON MISCONCEPTIONS

Students will have various types of prior knowledge about the concepts introduced in this lesson. Table 4.7 outlines some common misconceptions students may have concerning these concepts. Because of the breadth of students' experiences, it is not possible to anticipate every misconception that students may bring as they approach this lesson. Incorrect or inaccurate prior understanding of concepts can influence student learning in the future, however, so it is important to be alert to misconceptions such as those presented in the table.

Table 4.7. Common Misconceptions About the Concepts in Lesson 2

Topic	Student Misconception	Explanation
Energy	Energy is found only in living things.	Nonliving things also have energy, or the ability to do work. The composition of an object and its position determine the type of energy it has.
Maps*	Map symbols directly represent their referents.	Symbols may be abstract and may not be a true physical representation of the feature. For example, blue on a map does not always indicate water, a triangle does not always indicate a mountain, and green areas do not always indicate trees. The map legend provides a key for symbols.
Wind energy	Wind energy is expensive to produce.	Wind power is actually one of the lowest-priced renewable energy resources available, according to the DOE.

* For more information on misconceptions students might have about maps, see the National Geographic report "Spatial Thinking About Maps: Development of Concepts and Skills Across the Early Years" at *www.nationalgeographic.org/media/spatial-thinking-about-maps.*

PREPARATION FOR LESSON 2

Review the Teacher Background Information provided (p. 99), assemble the materials for the lesson, and preview the videos recommended in the Learning Plan Components section below. Download and have available topographic maps for your area and for the five regions of the United States. Be prepared to group students into teams of four in social studies class. These teams will continue to work together through the end of the module as they complete the Wind Farm Challenge.

LEARNING PLAN COMPONENTS
Introductory Activity/Engagement

Connection to the Challenge: Begin each day of this lesson by directing students' attention to the driving question for the module and challenge: Where could we locate a wind farm that a community would support? Hold a brief student discussion of how their learning in the previous days' lessons contributed to their ability to answer this question, using prompts such as "From what we learned about how wind is created, do you think there is the same amount of wind everywhere?" You may wish to hold a class discussion,

create a class list of key ideas on chart paper, or have students create a notebook entry with this information.

Social Studies Class: Begin the lesson by asking students to share their thoughts about how wind provides energy today. Ask them to consider the boats they created in the Blown Away Design Challenge and whether wind is an efficient source of direct energy. Ask if they think a flashlight could be powered by wind energy.

Show students the hand crank flashlight and demonstrate its operation to them. Ask students to share ideas about how wind could be used to power the flashlight. Guide students to understand that if you could attach windmill-type blades to the flashlight instead of the crank handle, the wind could power the flashlight. Tell students that a wind turbine works in essentially the same way as the flashlight works, by turning a magnet within a wire coil to create electricity.

Next, show the DOE video "Energy 101: Wind Turbines" (visit YouTube and search for this title or access the video directly at *www.youtube.com/watch?v=EYYHfMCw-FI*). Ask students what they noticed about the places where the wind turbines were located (e.g., no trees, flat areas, desert next to a mountain). Create a class list and ask students to look for common geographic features. Tell students that in this lesson, they are going to be grouped into teams for the Wind Farm Challenge, and each group will be assigned a geographic region of the United States.

Science Class: Ask students to offer ideas about how we know what we do about Earth's spheres (e.g., how do we know that temperature differences at Earth's surface cause wind?). Create a class list of their ideas. Focus students' attention on the role of scientists in learning about Earth. Ask students for their ideas about how scientists begin their investigations. Guide students to understand that scientists ask questions based on their observations or ideas and attempt to answer these questions through investigations and experiments.

Introduce students to the concept of scientific questioning by first offering a statement that starts with "I want to know," such as "I want to know if wind creates waves in the ocean." Tell students that scientists start their investigations by wanting to know about something, but then they change their "I want to know" statements to questions that can be answered. Repeat your "I want to know" statement, and ask students how that can be changed into a question ("Does wind create waves in the ocean?").

Introduce the idea that questions scientists ask should be testable by suggesting a variety of questions, including some that are testable (e.g., "What do we observe when wind blows on water?") and others that are not easily testable (e.g., "If we create more wind over the ocean, will the waves be larger?"). Next, pose this question to students: "Why do some people like pizza?" Ask students whether they think this question is testable (no). Then, ask them to think of a question that is similar but *is* testable (e.g.,

"Of the people in our class who like pizza, which part of the pizza do they like best?"). Ask students what the difference between these two questions is. Introduce the idea that testable questions have to do with scientific ideas rather than personal preferences or values, and that scientific questions can be tested through investigations (experiments, observations, or surveys).

Prepare a class chart with two columns, Testable Questions and Nontestable Questions, and have students prepare a similar chart in their STEM Research Notebooks. List attributes of testable questions in one column (e.g., they ask about living and nonliving things and events in the natural world; they are about scientific ideas; they can be tested or measured; we can collect data about them; they ask questions that we can answer with an investigation), and list attributes of nontestable questions in the other (e.g., they ask about opinions or values or things outside of our natural world; we can't collect evidence about them; we can't measure them; we might be able to research them for a science report, but we can't test them). Now, ask students to indicate whether the following questions are testable using the criteria in the charts you made:

- Why are people afraid of the dark? (no)

- Why do airplanes work? (no)

- How does changing the shape of a model airplane's wings affect its ability to fly? (yes)

- Why do plants grow? (no)

- How does changing the amount of water we give to a sunflower affect its growth? (yes)

Introduce the idea of variables to students, emphasizing that if many things change at once, it is difficult to tell what is causing other changes (e.g., if we observe the waves in the ocean when the wind starts to blow harder and it also begins to rain, how do we know what effect the wind has and what effect the rain has on changes in the waves?). Ask students for their ideas about how scientists contend with variables, creating a class list of student ideas. Introduce the terms *independent variable* as the variable that is changed or controlled by the scientists and *dependent variable* as the variable that might change in response to what the scientist does. Point out to students that testable questions always involve changing one variable (the independent variable) to see what happens to another variable (the dependent variable). Ask the students the questions above again, and have them identify whether each question includes independent and dependent variables, and if so, what the variables are. Then, have students complete the Testable or Not? Scientific Questions handout, and review their responses as a class.

Mathematics Connection: As a class, explore the weather graphs for your area on the National Weather Service website at *https://digital.weather.gov* by clicking on your location on the map displayed, then clicking on the thumbnails under Location Data in the pop-up to access larger graphs. Use these graphs to launch a discussion about labeling graph axes and measurement units such as degrees Fahrenheit and miles per hour (mph) by examining the various graphs available and asking students to identify the graph labels and the units on the axes.

ELA Connection: Have students create a notebook entry in which they describe and reflect on an experience with wind. Prompt students to use descriptive adjectives to give the reader an image of the wind as they experienced it. Students' experiences with wind will vary according to your geographic location, so you may wish to provide them with the following guiding questions or formulate questions that will best reflect your students' experiences:

- Have you stood on top of a treeless hill? How is this different from standing between buildings?

- Have you been on a sailboat?

- Have you ever seen a wind farm?

- Have you experienced a storm with high winds?

Activity/Exploration

Social Studies Class: Students explore the interaction of geography and wind via three activities: Marvelous Maps, Map Me!, and Where's the Wind?

Marvelous Maps

Show students a road map. Ask them if they can think of any other types of maps. Create a KWL chart, filling in the Know and Want to Know columns with student responses. Review the various types of maps with students, including topographic, relief, political, and weather maps. Review basic map terminology with students, including scale, map key or legend, and contour lines, as well as the use of symbols on maps. Examples of topographic and relief maps are shown in the student handouts.

Conduct a brief lesson in reading maps, using road maps and topographic maps (see Teacher Background Information section on p. 100 for details and link to USGS educational materials). Explain to students that topographic maps show landforms using contour lines and that the closer together the contour lines, the steeper the slope. Demonstrate measuring distances on a map using the map scale and a ruler.

Show the class a map of the five regions in the United States (see Teacher Background Information section on p. 100). Ask students to identify what region they live in and share what they believe are some features of this region that make it unique and distinguish it from other regions. Create a class list of students' ideas.

Now, divide students into the teams of four on which they will remain throughout the rest of the module to complete the Wind Farm Challenge. Have teams start their work by participating in a geography scavenger hunt. Distribute the U.S. Map Scavenger Hunt handouts to each student, and give each team a U.S. road map and rulers. Provide a reference map of the U.S. regions for the class. You may wish to give teams a time limit to finish.

After students compete this activity, return to the KWL chart, asking students to share what they learned about maps.

Map Me!

Tell students that each team will be assigned a region of the United States. The team will work together to identify a location for a wind turbine in this region for the Wind Farm Challenge. Show the class the reference map of U.S. regions and the topographic maps, and ask students if they think there is a region of the United States that is best for wind turbines. Hold a class discussion about what features might make a location good for a wind turbine (e.g., flat land, open space, offshore).

Assign each team a region. You may wish to do this by placing slips of paper with the regions' names in a paper bag and having teams choose their regions blindly. Have students work in their design teams to complete the Map Me! handout and blank map (attached at the end of this lesson), using colored pencils. Provide each team with a U.S. road map and access to the class reference map of U.S. regions.

Where's the Wind?

Introduce this activity by showing the video "Why Does the Wind Blow?" (visit YouTube and search for this title or access the video directly at *www.youtube.com/watch?v=xCLwbqmacck*). Tell students that for this activity, they should continue to work in their Wind Farm Challenge teams. Ask students if, based on what they saw in the video and what they know about wind, they think that different areas of the country have different amounts of wind resources.

Ask students to recall where wind farms are usually built (in places with predictable wind patterns). Ask them what else might influence decisions about wind farm locations. For example, there are strong winds at high elevations such as on mountains (however, the cost of building turbines where there is no easy road access and the lack of electrical transmission lines from the turbine to a power plant might pose problems). Ask students to brainstorm types of areas where wind farms might be built (e.g., desert, plains, prairie, the tops of rounded hills). Encourage students to consider that wind turbines can also be built offshore, in the ocean, and where winds are strong and predictable. Remind them, though that these options can cause the wind farm to become more expensive and more difficult to build the farther they are from the shore.

Students should have their maps from the Map Me! activity available for reference. Have student teams research the wind resources in their assigned regions using the maps on U.S. average wind speeds and U.S. offshore wind resources attached at the end of this lesson. You may also access state-level data for offshore wind resources at the DOE WINDExchange page on offshore wind maps at *https://apps2.eere.energy.gov/wind/windexchange/windmaps/offshore.asp*.

Next, have students access the interactive USGS wind farm map at *https://eerscmap.usgs.gov/windfarm/* to determine where wind farms are currently located in their team's region. After teams have researched their assigned regions, create a class database of wind energy potential by creating a chart and having students share details about the wind farms they identified. Your chart may look something like Table 4.8.

Table 4.8. Sample Table for Where's the Wind? Activity

Region	Range of average wind speeds	States with enough wind for a wind farm	States where wind farms already exist	Amount of energy a sample wind farm in this region produces
Northeast				
Southeast				
Midwest				
West				
Southwest				

Science Class: Students explore scientific questions and scientific tools used to measure natural phenomena in Earth's spheres. They then investigate wind speed measurement in the How Windy Is the Wind? activity. Remind students of the earlier discussion of testable questions (see science class in Introductory Activity/Engagement section on p. 103), and tell them that when they ask a testable question, they are starting to use the scientific method. Ask students to share what they know about the scientific method. Students should understand that this is the process that scientists use to answer a question. Introduce the steps of the scientific method to students:

1. Ask a question.

2. Form a hypothesis.

3. Test the hypothesis.

4. Analyze the results of the test.

5. Form a conclusion.

Write these steps on chart paper, leaving room under each to record additional information and student ideas. Have students compare the scientific method with the EDP introduced in Lesson 1 (see p. 54), asking students to share ideas about how the two are the same and how they are different. Ask them when the scientific method might be used instead of the EDP and vice versa, and when they might use both.

Ask students to share what they now know about scientific questions, adding these ideas under the "Ask a question" step on the chart paper. Pair students and have each pair create a question in their STEM Research Notebooks about drinking a caffeinated drink right before they go to bed (e.g., "Will drinking a caffeinated drink before bed change how much I sleep at night?"). Have each pair of students trade their question with another pair and discuss whether the question is testable. Review all the questions to ensure that they are testable. Note that students will not actually perform an experiment based on this question; this activity is designed only to guide students to think through the steps of the scientific process.

Now, focus students' attention on the second step of the scientific method. Ask students to share their ideas about what a hypothesis is (a statement of what you think the answer to your question is). Create a class definition and write this under the "Form a hypothesis" step on the chart paper. Students should understand that hypotheses are predictions based on observations or information they already have. Hypotheses can be phrased as "if, then" statements and should include independent and dependent variables. (e.g., "If I drink a caffeinated drink right before bed, then I will not sleep well"). Instruct student pairs to turn their questions into hypotheses and record the hypotheses in their STEM Research Notebooks. Remind students that they should be able to identify an independent variable (one they control) and a dependent variable (what might happen when the independent variable is changed) in their hypotheses. Tell students to circle the independent variable and underline the dependent variable.

Ask students to consider the next step in the scientific method, "Test the hypothesis." Have them share ideas about how scientists test their hypotheses, focusing on the idea that scientists create investigations to see whether and how the dependent variable changes when they change the independent variable. Ask students to share their ideas about how scientists know if a dependent variable changes. Introduce the idea that measurement is important in scientific investigations, and ask students to work in their teams to brainstorm about what changes scientists might measure (e.g., temperature, size, amount). Have each group share its list, and create a class list of ideas. Then, have students offer ideas about what tools scientists might use to measure each of these changes. As a class, review the testable questions on the Testable or Not? Scientific Questions handout, and have students identify what change they would measure for each question and how they would measure it.

Next, have student pairs revisit the hypotheses they created about drinking caffeinated drinks before bed and create a plan in their STEM Research Notebooks about how they would test the hypothesis using measurement and what tools they would use. Tell them that this plan should include a materials list (e.g., an 8 ounce caffeinated drink, a clock, a person to observe the sleeper), a numbered series of steps or a flow chart, and a table or graph representing how they would record data. Have each pair of students share their plan with another pair of students and give and receive feedback on the plans. Then, have each student pair share their plan with the class, focusing on how the plan would provide data that would answer the question. Give students time to revise their plans in response to the feedback they receive.

Now, have a class discussion on step four, "Analyze the results of the test," asking students how the data could be analyzed (e.g., making a table or graph to compare the amount of sleep the student gets on a night with no energy drink and on a night with the energy drink). Create a fictional set of data for the caffeinated drink investigation (e.g., when Sally drank an energy drink before bed, she slept for 5 hours; when she did not drink an energy drink before bed, she slept for 9 hours). Work as a class to create a sample graph for these data (a bar graph works best in this case), pointing out the x- and y-axes. Emphasize to students that one axis represents the independent variable, and the other the dependent variable.

Finally, discuss the last step of the scientific process, "Form a conclusion." Ask students what they are forming a conclusion about (the hypothesis) and how they will reach this conclusion (using the data they "collected" and analyzed). Emphasize to students that the conclusion should contain a statement about whether the hypothesis was or was not supported by the data they collected. Ask the class to consider how their conclusions might be influenced by outside factors (could something else besides the caffeinated drink have influenced the amount of sleep they got?). Discuss ways that these outside factors could be limited or controlled (e.g., repeating the measurements several times; making sure that the student got the usual amount of sleep the night before the test night; making sure that the noise level is the same each night).

Ask students what sphere they would be studying in their investigation about the student drinking the caffeinated drink (the biosphere). Ask them what other things in the biosphere scientists might measure. Have students work in their teams to complete the Measuring Earth's Spheres graphic organizer. Have each team share its ideas for one sphere about what could be measured and how these things could be measured. Hold a class discussion about scientific measuring tools, asking students to share ideas about what other measurement tools they know about. Show students a meter stick, a scale or balance, and a graduated cylinder or measuring cup. Ask the class whether these tools could be useful for measuring the wind, and if so, how they would be used. Tell students that there are specific tools used for measuring wind, and have them share ideas about why it might be important to measure the wind.

Ask if any students have experienced high winds before. Have the class name some weather events associated with high winds (e.g., hurricane, tornado, blizzard). Then, see if students can guess the highest wind speed ever recorded (about 250 miles per hour, in Australia during a typhoon or hurricane in 1996).

Ask students if they have heard the word *category* used to describe hurricanes. Tell them that categories are ratings of wind speed from 1 to 5. Hurricane wind categories using the Saffir-Simpson Hurricane Wind Scale are as follows:

- Category 1: 74–95 mph

- Category 2: 96–110 mph

- Category 3: 111–129 mph

- Category 4: 130–156 mph

- Category 5: 157 mph and higher

You may wish to show students the animation of increasing hurricane wind intensity provided by NOAA's National Hurricane Center at *www.nhc.noaa.gov/aboutsshws.php,* where you can also find descriptions of the types of damage caused by each category.

Then, show the video "Jim Cantore vs. Category 5 Winds," about experiencing hurricane force winds in a wind tunnel (visit YouTube and search for this title or access the video directly at *www.youtube.com/watch?v=pmJ8tXTcCfE*). Ask students to watch for the answers to the following questions, and discuss these with the class afterward:

- What is the highest wind Jim Cantore has experienced in an actual hurricane?

- What hurricane categories were the winds he experienced in the wind tunnel?

- What physical effects did he experience from the wind?

- What was the top wind speed he experienced in the wind tunnel?

How Windy Is the Wind?

Begin this activity by asking students how wind is described on weather reports (e.g., speed, direction, gusts). Have students share their ideas about how meteorologists collect this information. Show students the pictures of the anemometers in Figures 4.1 and 4.2 (see p. 102). Introduce the idea that an anemometer measures wind speed, and a weather vane measures direction. Ask students to look carefully at the cup anemometer and describe how they think it works. Have students identify the basic parts of the anemometer (e.g., cups, a rotating base for the cups). Have students brainstorm their ideas about how people can use an anemometer to determine how fast the wind blows. Create a class list of student ideas, guiding students to understand that counting the number of rotations of the cups within a certain amount of time will give them information about the wind's speed. Tell students that now they are going to work together in teams to create their own anemometers. Ask students which process—the EDP or the scientific method—they think they might use to design an anemometer. Tell them they will use the EDP to design and build their anemometers, but they will also use the scientific method as they test and redesign their devices.

Ask students what design features they think are important (must be able to turn freely, must stand upright, must catch the wind). Distribute the How Windy Is the Wind? handout, review the rules and materials for this activity with students. Remind students of the EDP steps and the class list of collaboration guidelines they created (students will be assessed on collaboration for this activity; see rubric attached at the end of this lesson).

Then, have student teams design, test, and redesign their anemometers. Students should complete How Windy Is the Wind? Engineer It! handouts, which ask about the steps of the EDP as well as the scientific method that students engage in. Have each group present its anemometer to the class, demonstrate how it works, and discuss the design process.

Mathematics Connection: Tell students that they are going to prepare a daily wind speed graph for their area that they will maintain throughout the remainder of the module. Hold a class discussion about what they should include on the graph (e.g., average wind speed, top wind speed, wind gusts). Ask students what sort of graph they think they should use for this (e.g., line graph, bar graph). Next, ask students what they think that the *x*- and *y*-axes of their graph should represent. Have students prepare their graphs. Then, as a class, access wind information for the current day and have students plot this on their graphs.

The National Renewable Energy Laboratory says that annual average wind speeds should be greater or equal to about 6.5 meters per second at an 80 meter height to be sufficient to power a wind turbine. Since students are likely to be most familiar with speeds in units of miles per hour and heights in feet and inches, work with the class to convert the speed and height.

ELA Connection: Continue exploration of the literature connections suggested in Lesson 1. Create a Reading Response section for students' STEM Research Notebooks. You may choose to use a variety of prompts based on the selected readings. This can be continued throughout the module with various pieces of literature.

Nonfiction examples include the following:

- What is the most important thing the author wants you to understand or know after reading this book? How do you know that?

- What surprised you the most in this book?

- How can what you learned in this book help you in the Wind Farm Challenge?

- What kinds of text features did this book use to help you learn? How did these affect your reading experience?

- Did this book remind you of another book, a movie, a real-life experience, or a current event? Why?

- What would you like to know more about after reading this book? Why? How could you find out more about this?

Fiction examples include the following:

- Would you have liked to live in the time and place in which this book was set? Why or why not?

- What character in this book would you choose to have as a friend? Why?

- What real-life events does this book remind you of? Why?

- Why do you think the author chose to write this story?

- From what you have read so far, what do you think will happen next?

Explanation

Social Studies Class: Introduce careers associated with mapping. These include geographers, cartographers, and photogrammetrists (see Teacher Background Information section on p. 100 for more information).

Science Class: Students should understand that anemometers are commonly used at weather stations to measure wind speed. Since an anemometer rotates at the same speed as the wind blows, wind speed is calculated by counting (usually electronically) the number of revolutions it makes in a specified period of time. This is usually converted to miles per hour for reporting to the public.

Have students copy the questions they identified as testable on the Testable or Not? Scientific Questions handout in their STEM Research Notebooks, leaving about half a page between questions. Ask students to identify which of Earth's spheres are involved in each question. Then, have students pose a hypothesis for each testable question. Ask students to choose one question with hypothesis and provide a plan (with a materials list and a series of steps or a flow chart) that shows how they would answer the question and collect data that would provide information to either support or refute their hypothesis.

Mathematics Connection: Since wind speeds are typically calculated as averages, this lesson presents an opportunity to introduce or review measures of central tendency (mean, median, mode) with students.

ELA Connection: Not applicable.

Elaboration/Application of Knowledge

Social Studies Class: Have students explore the NOAA Young Meteorologist Program interactive "Severe Weather Preparedness Adventure" game at *http://youngmeteorologist. org/game/index.html.*

Science Class: Have students complete the Earth's Spheres Assessment (attached at the end of this lesson) as a summative assessment of their understanding of the four spheres they investigated in this and the previous lesson.

An optional extension for this class is to have students conduct a deeper investigation into the effects of the interaction of geologic features and air masses on weather. One option is to have students visit the Smithsonian Science Education Center's interactive Weather Lab at *https://ssec.si.edu/weather-lab* to learn about the geologic and atmospheric causes of weather in North America.

STEM Research Notebook Prompt

Students should respond to the following prompt in their Research Notebooks: *You have been using the process that engineers use to solve problems: the engineering design process (EDP). Reflect on this process. How has it helped your team work together? What challenges do you face as you use the EDP with your team? Do you think that professional engineers have similar experiences working together to solve problems? Why or why not?*

Mathematics Connection: Have students create graphs to track the outside temperature during your class time each day of the module by reading an outdoor thermometer. Have students compare their temperature readings with the National Weather Service graph readings at *https://digital.weather.gov/* and offer explanations for any differences they observe.

An optional extension is to have students compile class data from all teams' anemometers and calculate average wind speeds based on trials from each group's device.

ELA Connection: Have students research current wind farm construction projects in the region where they live and write letters to their state representatives either expressing support for or opposition to the wind farm project.

Evaluation/Assessment

Students may be assessed on the following performance tasks and other measures listed.

Performance Tasks
- U.S. Map Scavenger Hunt handout
- Map Me! handout and blank map
- Where's the Wind? handout
- How Windy Is the Wind? handout
- How Windy Is the Wind? Engineer It! handout
- How Windy Is the Wind? Anemometer Design Rubric
- Earth's Spheres Assessment

Other Measures
- STEM Research Notebook Entry Rubric (see p. 88)
- Participation in Marvelous Maps activity
- How Windy Is the Wind? Collaboration Rubric

INTERNET RESOURCES

Map of U.S. regions
- *http://media.nationalgeographic.org/assets/file/us-regions-map.pdf*

USGS educational map-reading resources
- *http://education.usgs.gov/primary.html#geoggeneral*

Wind Energy Foundation FAQs
- *http://windenergyfoundation.org/about-wind-energy/faqs/*

U.S. Energy Information Energy Kids wind energy information
- *www.eia.gov/kids/energy.cfm?page=wind_home-basics*

Information about anemometers
- *www.nationalgeographic.org/encyclopedia/anemometer/*

Information about careers
- *www.bls.gov/ooh/life-physical-and-social-science/geographers.htm*

- *www.bls.gov/ooh/architecture-and-engineering/cartographers-and-photogrammetrists.htm*

- *www.wisegeek.com/what-are-photogrammetrists.htm*

- *www.bls.gov/ooh/life-physical-and-social-science/atmospheric-scientists-including-meteorologists.htm#tab-2*

Weather systems and patterns
- *www.noaa.gov/resource-collections/weather-systems-patterns*

National Geographic "Spatial Thinking About Maps: Development of Concepts and Skills Across the Early Years"
- *www.nationalgeographic.org/media/spatial-thinking-about-maps/*

"Energy 101: Wind Turbines" video
- *www.youtube.com/watch?v=EYYHfMCw-FI*

National Weather Service weather graphs
- *https://digital.weather.gov/*

"Why Does the Wind Blow?" video
- *www.youtube.com/watch?v=xCLwbqmacck*

Offshore wind maps
- *https://apps2.eere.energy.gov/wind/windexchange/windmaps/offshore.asp*

Interactive wind farm map
- *https://eerscmap.usgs.gov/windfarm/*

Saffir-Simpson Hurricane Wind Scale and animation of increasing wind intensity
- *www.nhc.noaa.gov/aboutsshws.php*

"Jim Cantore vs. Category 5 Winds" video
- *www.youtube.com/watch?v=pmJ8tXTcCfE*

Interactive "Severe Weather Preparedness Adventure" game

- *http://youngmeteorologist.org/game/index.html*

Interactive Weather Lab

- *https://ssec.si.edu/weather-lab*

Name: _____

STUDENT HANDOUT, PAGE 1

TESTABLE OR NOT? SCIENTIFIC QUESTIONS

For each question below, indicate whether the question is a scientific question by circling whether it is testable or not testable.

1. How does changing the amount of water given to a sunflower affect how it grows?

   Testable   Not testable

2. How do plants grow?

   Testable   Not testable

3. Who is the best singer?

   Testable   Not testable

4. Will more people like country music if we play it in school every day?

   Testable   Not testable

5. Why does water move?

   Testable   Not testable

6. Does stronger wind create taller waves in water?

   Testable   Not testable

Name: _____

STUDENT HANDOUT, PAGE 2

TESTABLE OR NOT? SCIENTIFIC QUESTIONS

7. Why does the Sun shine?

Testable   Not testable

8.   What is the best pet to have?

Testable   Not testable

9. Does changing the food you give your cat affect how much it eats?

Testable   Not testable

10. Will your goldfish be angry if you don't feed it?

Testable   Not testable

11. Are potato chips or corn chips better?

Testable   Not testable

Now, for each question you said is testable, circle the independent variable and underline the dependent variable.

Name: _____

STUDENT HANDOUT

MEASURING EARTH'S SPHERES

Earth's Spheres	What Is Measured?	What Tools Are Used?
Biosphere		
Lithosphere		
Hydrosphere		
Atmosphere		

Name: _____

STUDENT HANDOUT

U.S. MAP SCAVENGER HUNT

Work with your team using your U.S. map to answer the following questions as quickly as possible:

1. In which of the U.S. regions is California located? _____

2. Find the state of Maine. What states does it touch? _____

3. Find the state of Utah and count three states to the east. What state is this?

   _____

4. In what state(s) are the Rocky Mountains? _____

5. Is there a desert on your map? _____ What state is it in? _____

6. In what region is the Grand Canyon located? _____

7. Name the river that flows through Arkansas, Missouri, Tennessee, and Louisiana:

   _____

8. Using your map scale and ruler, calculate how many miles are between Newport News, Virginia, and San Francisco, California: _____

9. Name a mountain range in the Southeast region: _____

10. Name two states in the Midwest region: _____

11. What country does the Southwest region border? _____

12. How many miles are between Washington, DC, and Seattle, Washington? _____

13. Name the state in the Northeast region where the Declaration of Independence was signed (hint: the city where it was signed is Philadelphia): _____

14. What two countries does the United States border? _____

15. What states border the Pacific Ocean? _____

16. How many states border the Atlantic Ocean? _____

17. How many miles are between Boston, Massachusetts, and Dallas, Texas? _____

18. How many miles are between your school and Orlando, Florida? _____

19. What is the smallest state? _____ What region is it in? _____

20. What is the largest state? _____ What region is it in? _____

Name: _____

STUDENT HANDOUT, PAGE 1

MAP ME!

1. With your group, do the following:

   Use the road map to find the town or city where you live, and fill in the following information:

   I live in _____ (city or town),

   in _____ (county),

   in _____ (state).

   My state is in the _____ region of the United States.

2. Mark the location of your town on your blank map, and label your state, county, and town or city.

3. Now, identify the five regions of the United States, and shade those areas with different colors on your map.

4. Using topographic maps and road maps, identify the major geographic features of your assigned region (mountains, valleys, desert, large rivers) and mark those on your map, creating your own symbols for each feature.

5. Create a map key.

6. Label each state in your assigned region.

7. Put a star at the location of the capital city in each of the states in your assigned region.

4

Name: _____

My team's region: _____

STUDENT HANDOUT, PAGE 2

MAP ME!

Note: A full-color version of this image is available on the book's Extras page at www.nsta.org/roadmap-wind.

Wind Energy Lesson Plans

Name: _____   My team's region: _____

STUDENT HANDOUT

WHERE'S THE WIND?

1. The U.S. Department of Energy says that for an area to be suitable for a wind turbine, annual average wind speeds must be about 6.5 meters per second or greater. Use wind maps to answer the following questions:

    a. What is the highest average onshore wind speed in your region? _____

    b. What is the lowest average onshore wind speed in your region? _____

    c. If your region borders the ocean, what is the highest average offshore wind speed?

    _____

    d. If your region borders the ocean, what is the lowest average offshore wind speed?

    _____

    e. Are there areas in your region that have average wind speed high enough to build a wind

    turbine? _____

    f. What states in your region have enough wind for a wind turbine?

    _____

    _____

2. Using the interactive wind farm map at *https://eerscmap.usgs.gov/windfarm,* answer the following questions:

    a. Are there wind farms in your region? _____

    b. What states are the wind farms in? _____

    _____

3. Click on one of the farms in your region and answer the following questions (if the first wind farm you try doesn't have this information, try another one):

    a. What is the name of the farm? _____

    b. How much electricity can the farm generate (capacity)? _____

    c. What year did the farm start operating? _____

    d. How long are the turbine blades? _____

    e. How tall are the turbines? _____

STUDENT HANDOUT

U.S. AVERAGE WIND SPEEDS

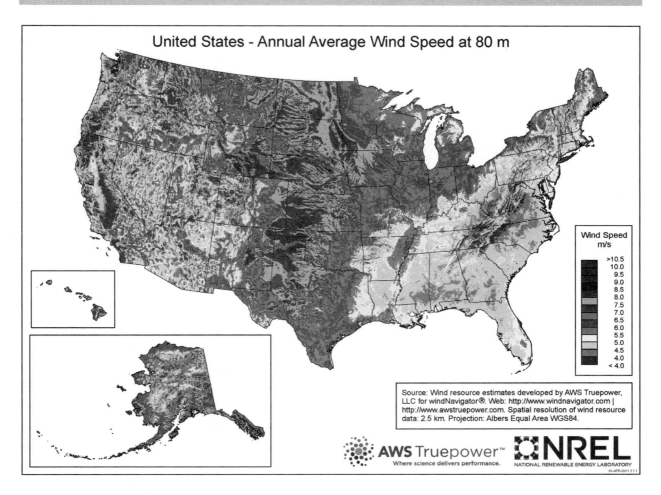

Source: National Renewable Energy Laboratories, U.S. Department of Energy. Public domain. *www.nrel.gov/gis/ images/80m_wind/USwind300dpe4-11.jpg.*

Note: A full-color version of this image is available on the book's Extras page at *www.nsta.org/roadmap-wind.*

STUDENT HANDOUT

U.S. OFFSHORE WIND RESOURCES

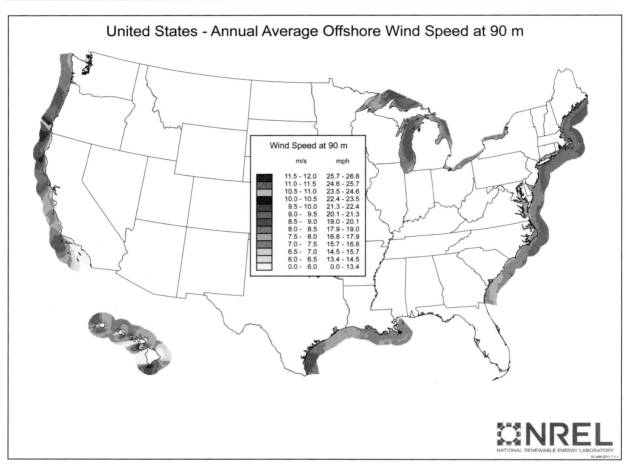

Source: National Renewable Energy Laboratories, U.S. Department of Energy. Public domain. *http://apps2.eere.energy. gov/wind/windexchange/windmaps/offshore.asp.*

Note: A full-color version of this image is available on the book's Extras page at *www.nsta.org/roadmap-wind.*

Name: _____ Team Name: _____

STUDENT HANDOUT

HOW WINDY IS THE WIND?

You and your team are scouting out possible sites for your wind farm. You traveled to the area yesterday, but when you packed up your gear to visit sites this morning, you realize that you left your anemometers behind! You are in a hurry and don't have time to buy special materials, so you must build your device using only the supplies you were able to find at the local grocery store.

Your challenge: Build a working anemometer to measure wind speed.

Rules

1. You must use only the materials supplied (you do not need to use all of them).

2. Your anemometer must move with the wind, without being touched by human hands. (Your teacher may provide a fan for wind if your class is unable to go outdoors.)

3. You must have a way to measure wind speed by its rotational rate (e.g., revolutions per minute).

4. All team members must participate in the design and building process.

5. You must demonstrate your device to the class and report the wind speed your team measured.

6. All members of your team must complete an Engineer It! handout.

Name: _____ Team Name: _____

ENGINEER IT! HOW WINDY IS THE WIND?

Step 1: (Define) State the problem (what are you trying to do?).

Step 2: (Learn) What solutions can you and your team imagine?

Step 3: (Plan) Sketch your design here and label materials. Be sure to consider safety issues when planning your design.

Step 4: (Try) Build it! Did everyone on your team have a chance to help?

Name: _____ Team Name: _____

_____
STUDENT HANDOUT, PAGE 2
_____

ENGINEER IT! HOW WINDY IS THE WIND?

Step 5: (Test) Test your design. How did it work? Could it work better?

How are you going to measure the wind with your anemometer?

Step 6: (Decide) Use the Scientifc Method!

Ask a question: What change might you make to your anemometer to make it work better?

_____

Form a hypothesis: If we _____, then

our anemometer will _____.

Test the hypothesis: Design an investigation to test your hypothesis. List the changes you will make to your anemometer and how you will determine how they worked:

STUDENT HANDOUT, PAGE 3

ENGINEER IT! HOW WINDY IS THE WIND?

Analyze the results of the test: Record your data here about what happened when you made your changes and tried your anemometer again:

Form a conclusion: Was your hypothesis correct, and how do you know if it was correct or incorrect?

_____

_____

_____

Step 7: Share your design! What features of your design will you point out to the class?

Name: _____

STUDENT HANDOUT, PAGE 1

EARTH'S SPHERES ASSESSMENT

1. Write these words below the name of the sphere that they belong to:

Tree	Lake	Fish	Dandelion	Mountain
Glacier	Lightning	Wind	Human	Desert

    a. Lithosphere

_____

    b. Atmosphere

_____

    c. Hydrosphere

_____

    d. Biosphere

_____

2. On the back of this sheet, draw and label a picture of a situation where three spheres interact. Write a sentence below describing how the spheres are interacting.

_____

_____

_____

Name: _____

STUDENT HANDOUT, PAGE 2

EARTH'S SPHERES ASSESSMENT

3. Under each description below, write the names of two Earth spheres that are interacting.

    a. Rain falls on the tomato plant you planted in your garden.

    _____

    b. A tree is blown down by hurricane winds.

    _____

    c. There is a flash flood, and creek water covers a road.

    _____

    d. A volcano erupts, and ash from it is in the air hundreds of miles away.

    _____

    e. You see a bird's nest blow out of a tree and land on your front porch.

    _____

How Windy Is the Wind? Collaboration Rubric (15 points possible)

Student Name: _____ Team Name: _____

Individual Performance	Below Standard (0–1)	Approaching Standard (2–3)	Meets or Exceeds Standard (4–5)	*Student Score*
INDIVIDUAL ACCOUNTABILITY	• Student is unprepared. • Student does not communicate with team members and does not manage tasks as agreed on by the team. • Student does not complete or participate in project tasks. • Student does not complete tasks on time. • Student does not use feedback from others to improve work.	• Student is usually prepared. • Student sometimes communicates with team members and manages tasks as agreed on by the team, but not consistently. • Student completes or participates in some project tasks but needs to be reminded. • Student completes most tasks on time. • Student sometimes uses feedback from others to improve work.	• Student is consistently prepared. • Student consistently communicates with team members and manage tasks as agreed on by the team. Student discusses and reflects on ideas with the team. • Student completes or participates in project tasks without being reminded. • Student completes tasks on time. • Student uses feedback from others to improve work.	
TEAM PARTICIPATION	• Student does not help the team solve problems; may interfere with teamwork. • Student does not express ideas clearly, pose relevant questions, or participate in group discussions. • Student does not give useful feedback to other team members. • Student does not volunteer to help others when needed.	• Student cooperates with the team but may not actively help solve problems. • Student sometimes expresses ideas, poses relevant questions, elaborates in response to questions, and participates in group discussions. • Student provides some feedback to team members. • Student sometimes volunteers to help others.	• Student helps the team solve problems and manage conflicts. • Student makes discussions effective by clearly expressing ideas, posing questions, and responding thoughtfully to team members' questions and perspectives. • Student gives useful feedback to others so they can improve their work. • Student volunteers to help others if needed.	

How Windy Is the Wind? Collaboration Rubric (*continued*)

Individual Performance	Below Standard (0–1)	Approaching Standard (2–3)	Meets or Exceeds Standard (4–5)	Student Score
PROFESSIONALISM AND RESPECT FOR TEAM MEMBERS	• Student is impolite or disrespectful to other team members. • Student does not acknowledge or respect others' ideas and perspectives.	• Student is usually polite and respectful to other team members. • Student usually acknowledges and respects others' ideas and perspectives.	• Student is consistently polite and respectful to other team members. • Student consistently acknowledges and respects others' ideas and perspectives.	

TOTAL SCORE: _____

Team Name _____

How Windy Is the Wind? Anemometer Design Rubric (20 points possible)

Team Performance	Below Standard (0–1)	Approaching Standard (2–3)	Meets or Exceeds Standard (4–5)	Team Score
CREATIVITY AND INNOVATION	• Design reflects little creativity with use of materials, lack of understanding of project purpose, and no innovative design features. • Design is impractical. • Design has several elements that do not fit.	• Design reflects some creativity with use of materials, a basic understanding of project purpose, and limited innovative design features. • Design is limited in practicality and function. • Design has some interesting elements but may be excessive or inappropriate.	• Design reflects creative use of materials, a sound understanding of project purpose, and distinct innovative design features. • Design is practical and functional. • Design is well crafted and includes interesting elements that are appropriate for the purpose.	
CONCEPTUAL UNDERSTANDING	• Design incorporates no or few features that reflect conceptual understanding of concepts.	• Design incorporates some features that reflect a limited conceptual understanding of concepts.	• Design incorporates several features that reflect a sound conceptual understanding of concepts.	
DESIGN REQUIREMENTS	• Design violates challenge rules or specifications. • Design is not finished.	• Design meets most challenge rules and specifications. • Design is finished on time.	• Design meets all challenge rules and specifications. • Design is finished on time.	
PERFORMANCE	• Device does not function or faces substantial problems in measuring wind speed in revolutions/minute.	• Device functions but may have some problems in measuring wind speed in revolutions/minute.	• Device functions and effectively measures wind speed in revolutions/minute.	

TOTAL SCORE: _____

Lesson Plan 3: Wind Impact

In this lesson, students investigate the economics of wind farms, their environmental impacts, and wind turbine blade design. Students consider the advantages and disadvantages of using wind energy, and then use the EDP and the scientific method to create and test models of various blade designs.

ESSENTIAL QUESTIONS

- How do the economic and environmental costs of using wind energy compare with the costs of using energy from fossil fuels?

- How does the design of a wind turbine's blades affect its efficiency in harnessing the wind's energy?

ESTABLISHED GOALS AND OBJECTIVES

At the conclusion of this lesson, students will be able to do the following:

- Research costs and income streams associated with wind farms and apply this information to create a budget

- Identify environmental impacts associated with wind farms

- Apply their understanding of the environmental impacts of wind farms to create a plan to mitigate one type of environmental disadvantage

- Use the EDP to design an innovation to mitigate an environmental impact of wind farms

- Collaborate with peers to create a solution to a problem

- Apply basic mathematical skills to understand average household energy usage and how various energy sources can meet consumer energy needs

- Identify the basic parts of a wind turbine and describe how a wind turbine functions

- Apply their understanding of wind turbine technology to build and test a simple wind turbine

- Apply their understanding of the EDP and the scientific process to create and test a variety of turbine blades

TIME REQUIRED

- 5 days (approximately 45 minutes each day; see Tables 3.8–3.9, p. 43)

MATERIALS

Required Materials for Lesson 3

- STEM Research Notebooks
- Internet access for student research and viewing videos
- Handouts (attached at the end of this lesson)

Additional Materials for Catch the Wind (1 per student pair unless otherwise indicated)

- Pinwheel pattern (attached at the end of this lesson)
- Cardstock
- Scissors
- Pencil
- Small electric motor (DC mini motor with output shaft; available at hobby stores)
- Piece of clay or cork
- 2 Alligator clip wires
- Box fan (1 per class)
- Voltmeter (1 per class)
- 2 Foam plates
- 18 inches heavy duty aluminum foil
- 6 craft sticks (regular size)
- 6 craft sticks (large size)
- 4 index cards (3 × 5 inches)
- 6 plastic drinking straws
- 1 roll of masking tape
- Ruler
- Catch the Wind handouts and data sheets (1 set per student)
- Safety glasses or goggles

Additional Materials for Dollars and Sense

- 15–20 boxes or bags

- "Receipts" for goods or services (small slips of paper with items and prices; 1 per student per good/service)

- Play money $10 bills (20 per student)

- Safety glasses or goggles

SAFETY NOTES

1. All laboratory occupants must wear safety glasses or googles during all phases of this inquiry activity.

2. Immediately wipe up any water on the floor to avoid a slip-and-fall hazard.

3. Use caution when operating electrical devices (e.g., fans) because of the potential shock hazard, especially near water.

4. Use caution when working with sharps (scissors, sticks, wires, alligator clips, fan blades, etc.) to avoid cutting or puncturing skin.

5. Make sure all materials are put away after completing the activity.

6. Wash hands with soap and water after completing the activity.

CONTENT STANDARDS AND KEY VOCABULARY

Table 4.9 lists the content standards from the *NGSS, CCSS,* and the Framework for 21st Century Learning that this lesson addresses, and Table 4.10 (p. 142) presents the key vocabulary. Vocabulary terms are provided for both teacher and student use. Teachers may choose to introduce some or all of the terms to students.

Table 4.9. Content Standards Addressed in STEM Road Map Module Lesson 3

NEXT GENERATION SCIENCE STANDARDS

PERFORMANCE EXPECTATIONS

- 3-5-ETS1-1. Define a simple design problem reflecting a need or a want that includes specified criteria for success and constraints on materials, time, or cost.

- 3-5-ETS1-2. Generate and compare multiple possible solutions to a problem based on how well each is likely to meet the criteria and constraints of the problem.

- 3-5-ETS1-3. Plan and carry out fair tests in which variables are controlled and failure points are considered to identify aspects of a model or prototype that can be improved.

Table 4.9. (*continued*)

DISCIPLINARY CORE IDEAS

ESS3.A. Natural Resources

- Energy and fuels that humans use are derived from natural sources, and their use affects the environment in multiple ways. Some resources are renewable over time, and others are not.

ETS1.A. Defining and Delimiting Engineering Problems

- Possible solutions to a problem are limited by available materials and resources (constraints). The success of a designed solution is determined by considering the desired features of a solution (criteria). Different proposals for solutions can be compared on the basis of how well each one meets the specified criteria for success or how well each takes the constraints into account. (3-5-ETS1-1)

CROSSCUTTING CONCEPTS

Energy and Matter

- Energy can be transferred in various ways and between objects.

Stability and Change

- Change is measured in terms of differences over time and may occur at different rates.

Cause and Effect

- Cause and effect relationships are routinely identified, tested, and used to explain change.

Influence of Science, Engineering, and Technology on Society and the Natural World

- Engineers improve existing technologies or develop new ones to increase their benefits, decrease known risks, and meet societal demands.

SCIENCE AND ENGINEERING PRACTICES

Asking Questions and Defining Problems

- Ask questions about what would happen if a variable is changed.
- Identify scientific (testable) and non-scientific (non-testable) questions.
- Ask questions that can be investigated and predict reasonable outcomes based on patterns such as cause and effect relationships.
- Use prior knowledge to describe problems that can be solved.
- Define a simple design problem that can be solved through the development of an object, tool, process, or system and includes several criteria for success and constraints on materials, time, or cost.

Planning and Carrying Out Investigations

- Plan and conduct an investigation collaboratively to produce data to serve as the basis for evidence, using fair tests in which variables are controlled and the number of trials considered.
- Make observations and/or measurements to produce data to serve as the basis for evidence for an explanation of a phenomenon or test a design solution.

Table 4.9. (*continued*)

Planning and Carrying Out Investigations (continued)

- Make predictions about what would happen if a variable changes.

- Test two different models of the same proposed object, tool, or process to determine which better meets criteria for success.

Analyzing and Interpreting Data

- Analyze and interpret data to make sense of phenomena, using logical reasoning, mathematics, and/or computation.

- Compare and contrast data collected by different groups in order to discuss similarities and differences in their findings.

- Analyze data to refine a problem statement or the design of a proposed object, tool, or process.

- Use data to evaluate and refine design solutions.

COMMON CORE STATE STANDARDS FOR MATHEMATICS

MATHEMATICAL PRACTICES

- MP1. Make sense of problems and persevere in solving them.

- MP2. Reason abstractly and quantitatively.

- MP3. Construct viable arguments and critique the reasoning of others.

- MP4. Model with mathematics.

- MP5. Use appropriate tools strategically.

MATHEMATICAL CONTENT

- 5.NBT.B.5. Fluently multiply multi-digit whole numbers using the standard algorithm.

COMMON CORE STATE STANDARDS FOR ENGLISH LANGUAGE ARTS

READING STANDARDS

- RI.5.1. Quote accurately from a text when explaining what the text says explicitly and when drawing inferences from the text.

- RI.5.4. Determine the meaning of general academic and domain-specific words and phrases in a text relevant to a grade 5 topic or subject area.

- RI.5.7. Draw on information from multiple print or digital sources, demonstrating the ability to locate an answer to a question quickly or to solve a problem efficiently.

- RI.5.9. Integrate information from several texts on the same topic in order to write or speak about the subject knowledgeably.

- RF.5.3. Know and apply grade-level phonics and word analysis skills in decoding words.

- RF.5.4. Read with sufficient accuracy and fluency to support comprehension.

- RF.5.4.a. Read grade-level text.

Table 4.9. (*continued*)

WRITING STANDARDS

- W.5.2. Write informative/explanatory texts to examine a topic and convey ideas and information clearly.

- W.5.4. Produce clear and coherent writing in which the development and organization are appropriate to task, purpose, and audience.

- W.5.6. With some guidance and support from adults, use technology, including the Internet, to produce and publish writing as well as to interact and collaborate with others; demonstrate sufficient command of keyboarding skills to type a minimum of two pages in a single sitting.

- W.5.7. Conduct short research projects that use several sources to build knowledge through investigation of different aspects of a topic.

- W.5.8. Recall relevant information from experiences or gather relevant information from print and digital sources; summarize or paraphrase information in notes and finished work, and provide a list of sources.

SPEAKING AND LISTENING STANDARDS

- SL.5.1. Engage effectively in a range of collaborative discussions (one-on-one, in groups, and teacher-led) with diverse partners on grade 5 topics and texts, building on others' ideas and expressing their own clearly.

- SL.5.1.b. Follow agreed-upon rules for discussions and carry out assigned roles.

- SL.5.1d. Review the key ideas expressed and draw conclusions in light of information and knowledge gained from the discussions.

- SL.5.6. Adapt speech to a variety of contexts and tasks, using formal English when appropriate to task and situation.

FRAMEWORK FOR 21ST CENTURY LEARNING

Interdisciplinary themes (global awareness; financial, economic, and business literacy; environmental literacy); Learning and Innovation Skills; Information, Media and Technology Skills; Life and Career Skills

Table 4.10. Key Vocabulary in Lesson 3

Key Vocabulary	Definition
budget	an estimate of expected expenses and available money over a period of time
energy	the ability to do work; a force that has the power to make things move or change
environmental factor	any action or condition that affects living things or communities
expenses	money spent during a given period of time to purchase goods or services or to pay for business costs
income	money received during a given period of time, including, for example, wages for an individual or sales dollars for a business
kilowatt-hour	a unit of energy that is equal to the work done by 1,000 watts for one hour
volts	a measure of the force with which electricity moves
watts	a measure of the work electricity does

TEACHER BACKGROUND INFORMATION

Since students will create a plan for a wind farm in response to their final challenge, it is important that they understand that scarcity of energy resources is not the only factor to be considered in using wind as an energy resource. This lesson therefore focuses on the environmental and economic costs of wind power. Students investigate various environmental impacts associated with wind energy production, learn about budgets and create a budget for their wind farm, and investigate wind turbine blade design.

Economic Costs of Wind Energy

Wind energy is considered one of the most economical forms of electricity, and its use is expected to expand over the coming years. Information on the costs associated with wind energy can be found at the American Wind Energy Association website at *www.awea.org/Resources/Content.aspx?ItemNumber=5547#CostofWindEnergy*.

Environmental Costs of Wind Energy

Although no fuel is directly used in generating wind energy, some environmental impacts have been associated with wind farms, including wildlife endangerment, noise pollution, and visual impacts. Some communities and environmental groups have protested wind energy based on these factors. An ABC news story about the protests surrounding the installation of wind turbines off the coast of Cape Cod illustrates the debates around these impacts (see the story at *http://abcnews.go.com/Technology/story?id=97849&page=1*).

COMMON MISCONCEPTIONS

Students will have various types of prior knowledge about the concepts introduced in this lesson. Table 4.11 outlines some common misconceptions students may have concerning these concepts. Because of the breadth of students' experiences, it is not possible to anticipate every misconception that students may bring as they approach this lesson. Incorrect or inaccurate prior understanding of concepts can influence student learning in the future, however, so it is important to be alert to misconceptions such as those presented in the table.

Table 4.11. Common Misconceptions About the Concepts in Lesson 3

Topic	Student Misconception	Explanation
Electricity	Electricity is created by generators or batteries.	Rather than creating electricity, generators or batteries cause electric charge to flow through wires.
Energy	Energy can be created or destroyed.	Energy cannot be created or destroyed but can be converted from one form to another (law of conservation of energy).
Wind energy	Wind energy is free.	Although the wind itself cannot be bought or sold and is freely available, the initial investment to build wind turbines is considerable.
Wind turbines	Individual wind turbines are a good way to provide power to people's homes.	Wind turbines require relatively large areas of clear land in windy areas; turbines located close to buildings and trees will not be efficient and therefore are not practical for many homeowners.

PREPARATION FOR LESSON 3

Review the Teacher Background information provided, assemble the materials for the lesson, and preview the videos recommended in the Learning Plan Components section below. For the Catch the Wind activity, copy the pinwheel pattern onto cardstock (one per student pair). The Dollars and Sense activity requires some preparation and setup

before students arrive; see this activity on page 147 in the Activity/Exploration section for more details.

LEARNING PLAN COMPONENTS
Introductory Activity/Engagement

Connection to the Challenge: Begin each day of this lesson by directing students' attention to the driving question for the module and challenge: Where could we locate a wind farm that a community would support? Hold a brief student discussion of how their learning in the previous days' lessons contributed to their ability to create their plan and build their prototype. You may wish to hold a class discussion, create a class list of key ideas on chart paper, or have students create a notebook entry with this information.

Social Studies Class: Hold a class discussion on the pros and cons of wind energy, creating a class list with two columns headed "Pros" and "Cons." Ask students to share their ideas about what might be benefits of using wind as an energy source. Write these in the "Pros" column of the class list. Now, ask students to consider what might be disadvantages of wind energy, and add their ideas to the "Cons" column of the class list. Show a video that provides a brief overview of the advantages and disadvantages of wind energy, called "Pros and Cons of Wind Power" (visit YouTube and search for this title or access the video directly at *www.youtube.com/watch?v=nlpD6BT0YuM*). Widely recognized benefits of wind energy are that it has no greenhouse gas emissions, it is the cheapest renewable energy source, and the land around turbines remains usable. Disadvantages are that it is less reliable than some other energy sources, and the turbines can harm bird and bat populations, can be loud, may be considered unsightly, and can block views.

After watching the video, have students add to the pros and cons on the class list. Ask students to think about this list and consider whether they believe we should increase the amount of wind energy we use in the United States. Take a class vote, and record student responses.

Science Class: Ask students to share their ideas about how wind turbines create electricity. Create a KWL chart and record students' ideas and questions. Prompt students to include information from the DOE video "Energy 101: Wind Turbines—2014 Update" video they watched in Lesson 2. You may wish to have students view this video a second time to watch for information about electricity production (visit YouTube and search for this title or access the video directly at *www.youtube.com/watch?v=EYYHfMCw-FI*). Add information to the Learned section of the KWL chart.

Mathematics Connection: Students will compare energy usage of various appliances in this lesson's activities. Although the mathematical concepts associated with energy usage are complex, students need only have a qualitative understanding that energy usage for homes is commonly measured in kilowatt-hours. Show students a utility bill that records electricity usage. Review the parts of the bill with students, and ask them to identify how much energy was used for that billing cycle. Ask them what units were used to measure electricity (kilowatt-hours). Show students the video "What Is a Kilowatt" for a brief explanation of kilowatts (visit YouTube and search for this title or access the video directly at *www.youtube.com/watch?v=lmIGonMm9jk*).

ELA Connection: Ask students to recall a time when the electricity went out at their homes or school. As a class, compile a list of adjectives describing the experience.

Activity/Exploration

Social Studies Class: Students consider the pros and cons of wind energy via an activity called Don't Bother the Neighbors. Introduce the activity by sharing with students that currently almost 6% of the United States' energy comes from wind power. Most of our electricity comes from fossil fuels such as coal and oil. Hold a class discussion on the pros and cons of using fossil fuels for energy. Compare this with the class list of pros and cons of wind energy. Tell students that there are plans to expand the use of wind energy in the United States. In this activity, they explore some of the environmental costs associated with wind energy. They will examine financial costs in math class. Before starting the activity, have each team come up with a name for its wind farm.

Don't Bother the Neighbors

Students work together in their teams to investigate the three major categories of environmental concerns associated with wind turbines and wind farms: wildlife endangerment, noise pollution, and visual impacts. Students should seek to understand the basis for these concerns, writing what they find in the boxes on the Don't Bother the Neighbors handout (attached at the end of this lesson).

After teams have completed their research into the environmental concerns, each team should then use the EDP to create a plan for a solution to address one of the environmental concerns they learned about. (You may wish to assign each team a category to ensure that all categories of concern are addressed.) Remind students that the EDP is used not only for creating products but also for solving problems. They might come up with a device or product (e.g., a device that keeps birds from coming near turbine blades), a business plan (e.g., buying the land in a radius around the farm), or another creative solution. Students should use the Don't Bother the Neighbors Engineer It! handouts to complete this task. (The "test" phase involves getting feedback from family members and peers.)

Then, each team should create a brief presentation (5 minutes or less) in which team members give an overview of the environmental concern their team is addressing and present a plan for addressing this concern. The goal of the presentation is to sell their

ideas to a group of fictional investors who are interested in investing in the wind farm. Students should use good presentation and collaboration practices (rubrics are attached at the end of this lesson) and should include visual aids in their presentations.

In the next lesson, student teams will create a model of the device or plan for the Wind Farm Challenge using simple materials that they bring from home. Prompt students at this time to think about what materials they will need, and have them create a plan for assembling those materials so that they will be on hand for their work in the Wind Farm Challenge.

Science Class: Students investigate wind turbines and blade design in the Catch the Wind activity. Introduce the activity by showing the video "How Wind Turbines Generate Electricity" at *http://fwee.org/nw-hydro-tours/how-wind-turbines-generate-electricity/*. Tell students that now they will create a simple wind turbine in the form of a pinwheel. In part 1 of this activity, students will use voltmeters to measure the voltage produced by their pinwheels. A simple way to explain voltage is to compare it to water moving through a hose. Just as pressure is needed to make water flow through a hose, pressure is also needed to make electric current flow through wires. We call this electric pressure *voltage*, and we use volts to measure the electric pressure. Batteries or generators (such as wind turbines) supply voltage. In part 2 of the activity, students will use the scientific method to experiment with a variety of blade materials and lengths to see how changing these variables affects their turbine's performance.

Catch the Wind

Have students work in pairs to create pinwheels connected to a small motor, and then measure the voltage to test the electric force the wind creates when it turns the pinwheels (see Catch the Wind student handouts attached at the end of this lesson for detailed instructions). Next, student teams use the scientific process to test various blade lengths and blade materials. Have the teams formulate four separate hypotheses. They should then test their hypotheses by changing one variable at a time—either blade material or blade length—and measuring the voltage produced by the redesigned pinwheel to test their hypotheses. Be sure that for each test, students place their pinwheels at a constant distance from the fan (12 inches) and use the same fan setting. It may be helpful to place a piece of masking tape 12 inches from the front of the fan to indicate the position at which the pinwheels should be tested.

Mathematics Connection: Students investigate budgeting in the Dollars and Sense activity, create a budget for their wind farm in the Dollars and Wind activity, and explore energy usage via the Energy Explorers activity.

Dollars and Sense

Set up 15 to 20 boxes or bags around the room labeled to represent various goods and services, and place a pile of "receipts" (enough for each student in the class) in front of each with the name of the item or service purchased and the cost. Be sure that some goods and services represent wants and others needs, and put them in random order, with wants and needs mixed together. Following are some examples of goods and services to include:

- Housing: $100

- Food: $60

- Haircut or style: $10

- Manicure: $10

- iPhone bill: $20

- Satellite TV: $20

- Car: $100

- Kitten or puppy: $10

- Savings account: $10

- Meal at a restaurant: $10

- New clothes: $20

- Gas for car to get to work: $10

- Shoes (name whatever brand is popular with your students): $20

- Health insurance: $20

- New Kindle book: $10

- Trip to the beach: $100

- Gym membership: $20

Give each student 20 $10 bills in play money. Tell them that they have been hired by a wind energy company, and this is their salary for the week. Give students 5 to 10 minutes to go around the room and put money into each box that represents an item or service they wish to "purchase," taking a receipt in exchange.

After the purchasing is complete, have students add their receipts to make sure that they spent their $200 and no more. If any students exceeded $200, they should give up goods or services so that their total is $200. If they are short of $200, they may take additional receipts.

Now, read through the list one item at a time, having students stand up whenever an item they purchased is called out. Save the necessities, such as food, housing, and gas for work, until last. Note whether all students stand up for the necessities, and ask students why they think everyone did or did not spend money on the necessities.

Dollars and Wind

Students work together in their challenge teams for this activity to create budgets for their fictional wind farms using the Dollars and Wind handout (attached at the end of this lesson). A list of "facts" for students to use in creating their budgets is provided on page 1 of the handout (p. 153).

Energy Explorers

Students investigate the concept of kilowatt-hours and energy usage in this activity. Have students work in pairs or their teams to create a list of appliances in their homes that use electricity and the amount of time they estimate that each appliance is used each week. Students should focus on large appliances and small electronics that are on all the time. Next, have students use the DOE's appliance energy calculator at *https://energy.gov/energysaver/estimating-appliance-and-home-electronic-energy-use* for the appliances on their lists.

Have students create a table for data and calculate total weekly appliance energy consumption based on the appliances and usage rates they chose. Compile all totals for the class, and have students calculate an average weekly appliance energy consumption for the class. Investigate data from pairs or teams, looking for trends in energy usage in the class and identifying particularly high usage and high energy consuming appliances.

Point out to students that the energy calculator requires watts but that energy consumption is usually measured in kilowatts. Have students practice converting watts (W) to kilowatts (kW) using the following formula: $kW = W/1,000$. Point out to students that the kilowatt-hours on the electric bill they saw in mathematics class earlier (see p. 144) is kilowatts multiplied by a number of hours.

ELA Connection: Have students create a STEM Research Notebook entry titled "A Day Without Electricity," in which they use creative writing to tell the story of a day when electricity has suddenly become unavailable.

Explanation

Social Studies Class: Students should understand that all sources of energy have environmental consequences. You may wish to introduce the greenhouse effect to illustrate one of the environmental effects of burning fossil fuels. Explain to students that the greenhouse effect happens when certain gases collect in Earth's atmosphere. Greenhouse gases include carbon dioxide, methane, nitrous oxide, fluorinated gases, and ozone. These gases allow the Sun's light to shine onto Earth, but they trap some of the heat that is reflected back into the atmosphere. This is important because it allows Earth to be warm enough to sustain life; without the greenhouse effect scientists estimate that the average temperature of Earth would drop from 57°F to about 0°F. Since the early 1800s, however, people have been releasing large amounts of greenhouse gases into the atmosphere, and scientists believe that this is increasing the greenhouse effect, which increases Earth's temperature. The largest source of greenhouse gas emissions in the United States is from burning fossil fuels for electricity, heat, and transportation.

Science Class: Students need to have a basic understanding of energy as a property of objects that can be transferred to other objects or converted into different forms (heat, for example). Students can see evidence of energy around them every day and may particularly associate energy with electricity. They should understand that electricity carries energy and that it takes a source of energy (e.g., coal, solar power, wind power) to create the electricity that is delivered to our homes and schools.

Whereas voltage is the force that moves electricity through a wire, electric current is a measure of the amount of electricity that can flow through a material. This amount is measured in amperes, or amps. To measure amps, how much electric current flows through a wire in a certain amount of time is measured, similar to measuring how much water flows through a hose in a given amount of time.

Have students create a STEM Research Notebook entry in which they reflect on their pinwheel designs. Students should include the following:

- How the material used in their pinwheel blades influenced the voltage reading and why they think this might be.

- How the length of the pinwheel blades influenced the voltage reading and why they think this might be.

- How, if they were wind turbine designers, they might use this information to create wind turbine blades.

Mathematics Connection: Students should understand the budgeting concepts introduced in the Dollars and Sense activity. In particular, students should understand the concept of income and expenditures and that a balanced budget means that expenditures in a given time are no greater than income in that same period of time. You may wish to demonstrate this to students by creating a simple budget table with two columns, one for income and one for expenses. Be sure to designate a monthly income and a list of monthly expenses. Hold a class discussion about why keeping a budget might be useful for an individual and why keeping a budget might be useful for a business.

ELA Connection: Hold a class discussion about what creates a good narrative text. Provide students with the following tips about narrative text:

- It tells a story rather than just presenting a series of events.

- The story typically has a central idea or problem to be resolved.

- The writer is clear about why he or she is writing the story and focuses on a specific event.

- The writer includes details relevant to the story.

- The story is organized.

- The writer shows rather than tells.

- Dialogue is used to advance the story

Elaboration/Application of Knowledge

Social Studies Class: Hold a class discussion asking students to compare and contrast wind energy with another form of renewable energy such as solar power. Ask them to consider costs, availability, and environmental factors as they research and compare the two energy sources. Create a class list of student ideas.

STEM Research Notebook Prompt

Students should respond to the following prompt in their Research Notebooks: Imagine that you are a cattle farmer living on 45 acres in the country, where your cattle graze. You learn that a wind energy company is going to install a wind farm with 100 wind turbines on the property next door to you. How do you feel about this? Write a letter to the company either expressing your support for the project or expressing your objection to the wind farm. Provide at least two reasons for your position, and be as specific as possible.

Science Class: Have students complete the Scientific Method Assessment (attached at the end of this lesson). Then, have students share their best blade designs from the Catch the Wind activity with the class. Compare these blade designs, identifying features they share and points of difference. Have students create a labeled sketch in their STEM Research Notebooks of a pinwheel design that they think combines the best attributes of the pinwheels they have seen, and then reflect on what features they believe made these pinwheels work well.

Mathematics Connection: Not applicable.

ELA Connection: Have students exchange the letters they created in their STEM Research Notebook prompt, and have students respond to their classmates' letters. Discuss the

format of a business letter and how the language used in a business letter might be different from that used in a narrative text.

Evaluation/Assessment

Students may be assessed on the following performance tasks and other measures listed.

Performance Tasks

- Dollars and Wind handout and budget sheet
- Don't Bother the Neighbors handout
- Don't Bother the Neighbors Engineer It! handout
- Don't Bother the Neighbors Presentation Rubric
- Catch the Wind student handouts
- Scientific Method Assessment

Other Measures

- STEM Research Notebook Entry Rubric (see p. 88)
- Participation in class discussions
- Don't Bother the Neighbors Collaboration Rubric

INTERNET RESOURCES

Costs associated with wind energy

- *www.awea.org/Resources/Content.aspx?ItemNumber=5547#CostofWindEnergy*

ABC news story on protests over installation of wind turbines

- *http://abcnews.go.com/Technology/story?id=97849&page=1*

"Pros and Cons of Wind Power" video

- *www.youtube.com/watch?v=nlpD6BT0YuM*

"Energy 101: Wind Turbines—2014 Update" video

- *www.youtube.com/watch?v=EYYHfMCw-FI*

"What Is a Kilowatt" video

- *www.youtube.com/watch?v=lmIGonMm9jk*

"How Wind Turbines Generate Electricity" video
- *http://fwee.org/nw-hydro-tours/how-wind-turbines-generate-electricity*

Appliance energy calculator
- *https://energy.gov/energysaver/estimating-appliance-and-home-electronic-energy-use*

4

Name: _____   Wind Farm Name: _____

DOLLARS AND WIND

Your team is conducting an analysis to see how cost-efficient it will be to build a wind farm. Answer the questions on this page, and then use your answers to prepare the Dollars and Wind budget on page 2.

You know the following facts:

- ✓ Land costs $1,000 per acre.

- ✓ Each turbine requires ½ acre of land.

- ✓ You can lease land around your turbines to a local farmer, who will pay you $200 per acre to farm this land.

- ✓ You will provide electricity to 33,000 households.

- ✓ The substation where the electricity from your wind farm is sent is 4 miles from your wind farm.

- ✓ Cable to carry electricity to the substation costs $5,000 per mile.

- ✓ A turbine costs $1,000,000 per megawatt (MW).

- ✓ The kind of wind turbine you will use produces 1.5 MW each hour (1.5 megawatt-hours, or MWh), so it has a 1.5 MW capacity.

- ✓ One turbine can power 330 homes.

- ✓ Each home uses 11 MWh per year.

- ✓ Consumers pay $65 per MWh for their electricity.

- ✓ It costs $50 to produce 1 MWh of wind energy.

- ✓ The government is providing you with a $3,000,000 grant to start your project.

Before you create your budget, answer the following questions:

1. How many turbines will you need in order to power all 33,000 homes? (Hint: You know that 1 turbine can power 330 homes, so 330 × **?** = 33,000.) _____

2. How many acres of land do you need for your turbines? (Hint: Each turbine requires **½** acre of land.) _____

3. How many MWh will the 33,000 households use each year? (Hint: each household uses 11 MWh each year.) _____

Wind Energy, Grade 5

Name: _____   Wind Farm Name: _____

STUDENT HANDOUT, PAGE 2

DOLLARS AND WIND

Using your answers to the questions on page 1 and the information you found from your research, create a budget for the first year of operation for your wind farm, including all sources of income and expenses. Then use your totals to calculate the difference and see which is greater.

Dollars and Wind Budget for _____ (name of your wind farm)

Number of Turbines: _____   Acres: _____

Income		
Source of Income	**Calculations**	**Amount of Income**
Selling electricity to 33,0000 households		
Leasing land to farmer		
Government grant		
Total income		

Expenses		
Type of Expense	**Calculations**	**Amount of Expense**
Land		
Turbines		
Installing cable to substation		
Cost to produce electricity for 330 homes		
Total expenses		

Difference
Total income − Total expenses =

Name: _____

STUDENT HANDOUT, PAGE 3

DOLLARS AND WIND

After you have completed your budget, answer the following questions:

What was your total income? _____

What was the total of your expenses? _____

Were expenses more than income? How much more? _____

_____

Wind Energy Lesson Plans

Name: _____   Wind Farm Name: _____

STUDENT HANDOUT

DON'T BOTHER THE NEIGHBORS

You've gotten some good news! A group of investors is interested in investing $500,000 in your wind farm. The bad news is that the investors have learned that there are concerns in the surrounding community about the environmental impact of the farm. You have a meeting set up with the investors to talk about this. To prepare for this meeting, you need to research the environmental concerns to understand what the impacts of the wind farm are and why they might be a problem, and then devise a plan to address one of the concerns, using the Engineer It! handouts. You will present this plan to the investors to persuade them to invest in your farm.

In the boxes below, describe each concern related to wind farms and why it is a problem. Give your opinion on whether it is a serious problem. Why or why not? List one idea for addressing each problem.

Wildlife Endangerment

Noise Pollution

Visual Impacts

Name: _____   Wind Farm Name: _____

STUDENT HANDOUT, PAGE 1

ENGINEER IT! DON'T BOTHER THE NEIGHBORS

Step 1: (Define)   State the problem (what are you trying to do?).

Step 2: (Learn)   What solutions can you and your team imagine?

Step 3: (Plan)   What solution does your team think is best? What do you need to consider as you create your plan? If you are designing a device or product, sketch your design here. Be sure to consider safety issues when planning your design.

Step 4: (Try)   List the steps in your plan. What will you need to do to make your plan a reality? If you are designing a device, what will you need to do to build it? You will build a model of this in the final challenge!

Name: _____   Wind Farm Name: _____

STUDENT HANDOUT, PAGE 2

ENGINEER IT! DON'T BOTHER THE NEIGHBORS

Step 5: (Test) Present your plan to another team or to family members. Do they have any suggestions to make it better?

Step 6: (Decide) Will you change anything in your plan? What?

Step 7: Share your plan! How will you present this plan to your investors so that they want to invest in your farm? Think about what each team member will do.

Name: _____

STUDENT HANDOUT, PAGE 1

CATCH THE WIND

There are two parts to this activity. In part 1, you and your partner will create a mini wind turbine using a pinwheel pattern and determine how much electric force it generates. In part 2, you will use the scientific method to improve on the mini wind turbine design.

Part 1 Instructions

1. To create your pinwheel, first cut out the pattern along the solid lines.

2. Cut dotted lines just to the edge of the circle in the middle (avoid cutting into the center circle).

3. Use a pencil to punch holes in the dots on the four corners and in the center of the pattern where the lines meet.

4. Fold the sections of the pinwheel in so that the small circles on the corners meet the center circle.

5. Put the shaft of the motor through the center of the pinwheel.

6. Put the piece of clay or cork on the end of the shaft to keep the pinwheel in place.

7. Place your turbine in front of the fan to make sure that it will spin.

8. Attach the alligator clips to the motor and the voltmeter, and place your turbine 12 inches from the fan.

9. Record the number of volts (electric force) from the voltmeter reading on Data Sheet, Page 1.

Name: _____

CATCH THE WIND

In this part of the activity, you will create and test new blade designs on your mini wind turbine. You will create new blades with different materials and different blade lengths to see how these changes affect your wind turbine performance.

Possible Blade Materials

Foam plate

Heavy duty aluminum foil

Craft sticks (regular and large size)

3 × 5 inch index cards

Plastic drinking straws

Part 2 Instructions

1. Remove the pinwheel from the motor.

2. Look at the materials listed above to create a hypothesis about how to make a more efficient mini wind turbine (generate more electric force).

3. Record your hypothesis on page 2 of your data sheet

4. Create a new set of turbine blades based on your hypothesis, and attach those to the motor.

5. Record the blade materials and lengths in the second column of the chart on page 1 of your data sheet.

6. Place the turbine 12 inches in front of the fan, and measure the electricity output using the voltmeter. Record this in the third column of the chart on page 1 of your data sheet.

7. Repeat the above procedure three more times, changing one independent variable (blade material or blade length) each time and recording everything on your data sheet.

8. Answer the questions on page 2 of the data sheet.

Name: _____

DATA SHEET, PAGE 1

CATCH THE WIND

Data Item	Independent Variable: Blade Material and Length	Dependent Variable: Electric Force (Volts)
Pinwheel pattern		
Hypothesis 1		
Hypothesis 2		
Hypothesis 3		
Hypothesis 4		

Name: _____

DATA SHEET, PAGE 2

CATCH THE WIND

Hypothesis 1:

_____

_____

Hypothesis 2:

_____

_____

Hypothesis 3:

_____

_____

Hypothesis 4:

_____

_____

1. What material worked best? _____

2. What materials didn't work well? _____

3. Did the length of the turbine blades affect the amount of electric force? If so, how?

_____

4. What are your conclusions?

_____

_____

STUDENT HANDOUT

PINWHEEL PATTERN FOR CATCH THE WIND

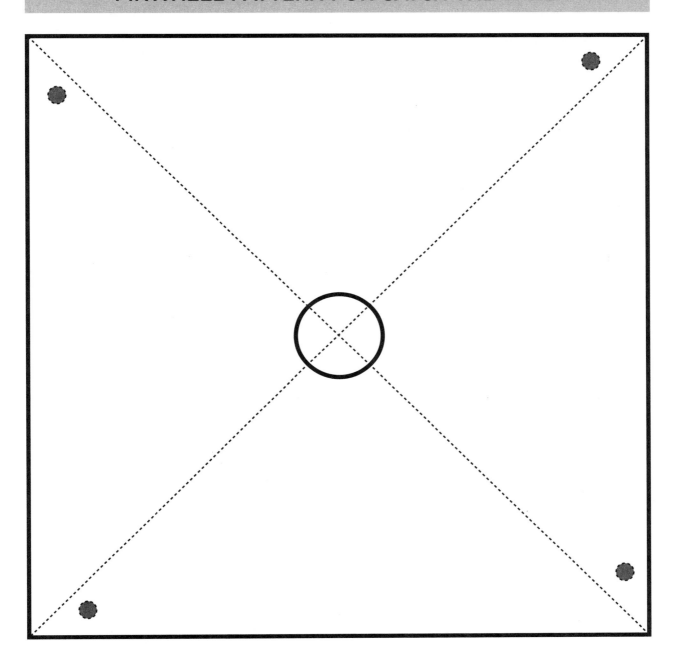

SCIENTIFIC METHOD ASSESSMENT

1. Which of the following is an example of a question that is testable with a scientific investigation?

    a. Why do some people like the color purple?
    b. Who made the first anemometer?
    c. Does the amount of sunlight shining on a rock change the temperature of the rock?
    d. How do apples grow?

2. The correct order of the scientific process is to form a hypothesis, test the hypothesis, ask a question, analyze the results of the test, form a conclusion.

    a. True
    b. False

3. Something that can be changed or controlled in a scientific investigation is called

    _____.

    a. A hypothesis
    b. A variable
    c. An exponent
    d. An experiment

4. In an experiment, scientists should change many variables at the same time.

    a. True
    b. False

Name: _____

SCIENTIFIC METHOD ASSESSMENT

5. Which of the following is an example of data?

    a. Water temperature
    b. Number of birds in an oak tree
    c. How many people in our class like ice cream
    d. All of the above

6. Which of these is a conclusion, using the scientific method?

    a. Tree leaves are green.
    b. Trees have bark.
    c. The tree is 40 feet tall.
    d. A sunflower that is watered every day grows faster than a tree that is watered once a week.

7. Which of the following is a tool scientists use to measure the speed of the wind?

    a. Speedometer
    b. Balance
    c. Weather vane
    d. Anemometer

Name: _____

STUDENT HANDOUT, PAGE 3

SCIENTIFIC METHOD ASSESSMENT

8. If your hypothesis is "If a sunflower seed is placed in the sun, then it will sprout more quickly than one in the dark," which of the following is the independent variable?

   a. The amount of water you give the seed
   b. How much sunlight there is on the day you observe the seed sprouting
   c. The amount of sunlight the sunflower seeds are exposed to after you plant the seeds
   d. How fast the seeds sprout

9. In the following statement, circle the independent variable and underline the dependent variable:

   If I leave the lights on in my house all day, our family's electric bill will be higher.

10. Creating a graph is an example of a way to _____

   _____.

   a. Analyze data
   b. Create an investigation
   c. Draw a conclusion
   d. Ask a testable question

Don't Bother the Neighbors Collaboration Rubric (15 points possible)

Student Name: _____   Team Name: _____

	Below Standard (0–1)	Approaching Standard (2–3)	Meets or Exceeds Standard (4–5)	Student Score
Individual Performance **INDIVIDUAL ACCOUNTABILITY**	• Student is unprepared. • Student does not communicate with team members and does not manage tasks as agreed on by the team. • Student does not complete or participate in project tasks. • Student does not complete tasks on time. • Student does not use feedback from others to improve work.	• Student is usually prepared. • Student sometimes communicates with team members and manages tasks as agreed on by the team, but not consistently. • Student completes or participates in some project tasks but needs to be reminded. • Student completes most tasks on time. • Student sometimes uses feedback from others to improve work.	• Student is consistently prepared. • Student consistently communicates with team members and manage tasks as agreed on by the team. Student discusses and reflects on ideas with the team. • Student completes or participates in project tasks without being reminded. • Student completes tasks on time. • Student uses feedback from others to improve work.	
TEAM PARTICIPATION	• Student does not help the team solve problems; may interfere with teamwork. • Student does not express ideas clearly, pose relevant questions, or participate in group discussions. • Student does not give useful feedback to other team members. • Student does not volunteer to help others when needed.	• Student cooperates with the team but may not actively help solve problems. • Student sometimes expresses ideas, poses relevant questions, elaborates in response to questions, and participates in group discussions. • Student provides some feedback to team members. • Student sometimes volunteers to help others.	• Student helps the team solve problems and manage conflicts. • Student makes discussions effective by clearly expressing ideas, posing questions, and responding thoughtfully to team members' questions and perspectives. • Student gives useful feedback to others so they can improve their work. • Student volunteers to help others if needed.	

Don't Bother the Neighbors Collaboration Rubric (*continued*)

Individual Performance	Below Standard (0–1)	Approaching Standard (2–3)	Meets or Exceeds Standard (4–5)	Student Score
PROFESSIONALISM AND RESPECT FOR TEAM MEMBERS	• Student is impolite or disrespectful to other team members. • Student does not acknowledge or respect others' ideas and perspectives.	• Student is usually polite and respectful to other team members. • Student usually acknowledges and respects others' ideas and perspectives.	• Student is consistently polite and respectful to other team members. • Student consistently acknowledges and respects others' ideas and perspectives.	

TOTAL SCORE: _____

Team Name: _____

Don't Bother the Neighbors Presentation Rubric (30 points possible)

Team Performance	Below Standard (0–2)	Approaching Standard (3–4)	Meets or Exceeds Standard (5–6)	Team Score
INFORMATION	• Team includes little interesting or factual information and may omit important information.	• Team includes some interesting and factual information but could provide more details.	• Team includes complete, interesting, and factual information.	
IDEAS AND ORGANIZATION	• Team does not have a main idea or organizational strategy. • Presentation does not include an introduction or conclusion. • Presentation is confusing and uninformative. • Team uses presentation time poorly and it is too short or too long.	• Team has a main idea and organizational strategy, although it may not be clear. • Presentation includes either an introduction or conclusion or both. • Presentation is fairly coherent, well organized, and informative. • Team uses presentation time adequately, but presentation may be slightly too short or too long.	• Team has a clear main idea and organizational strategy. • Presentation includes both an introduction and conclusion. • Presentation is coherent, well organized, and informative. • Team uses presentation time well and presentation is an appropriate length.	
PRESENTATION STYLE	• Only one or two team members participate in the presentation. • Presenters are difficult to understand. • Presenters use language inappropriate for audience (slang, poor grammar, frequent filler words such as "uh," "um").	• Some, but not all, team members participate in the presentation. • Most presenters are understandable, but volume may be too low or some presenters may mumble. • Presenters use some language inappropriate for audience (slang or poor grammar, some use of filler words such as "uh," "um").	• All team members participate in the presentation. • Presenters are easy to understand. • Presenters use appropriate language for audience (no slang or poor grammar, infrequent use of filler words such as "uh," "um").	

Don't Bother the Neighbors Presentation Rubric (*continued*)

Team Performance	Below Standard (0–2)	Approaching Standard (3–4)	Meets or Exceeds Standard (5–6)	Team Score
VISUAL AIDS	• Team does not use any visual aids to presentation. • Visual aids are used but do not add to the presentation.	• Team uses some visual aids to presentation, but they may be poorly executed or distract from the presentation.	• Team uses well-produced visual aids or media that clarify and enhance presentation.	
RESPONSE TO AUDIENCE QUESTIONS	• Team fails to respond to questions from audience or responds inappropriately.	• Team responds appropriately to audience questions, but responses may be brief, incomplete, or unclear.	• Team responds clearly and in detail to audience questions and seeks clarification of questions.	

TOTAL SCORE: _____

Lesson Plan 4: The Wind Farm Challenge

In this lesson, students synthesize their learning from the previous lessons as each team creates a proposal for a wind farm in a location of its choice within the team's assigned region of the United States. The proposal should be designed to convince the local community that the wind farm is a desirable addition. In its proposal, each student team should consider the proposed site's landforms and historical weather patterns, as well as local economic factors and the wind farm's potential to produce energy. Additionally, students address environmental impacts and propose mitigation plans for potential negative impacts, providing a model of this mitigation plan.

ESSENTIAL QUESTION

- Where could we locate a wind farm that a community would support?

ESTABLISHED GOALS AND OBJECTIVES

At the conclusion of this lesson, students will be able to do the following:

- Apply their understanding of economic, environmental, and technological features of wind turbines and wind farms to create a proposal for a wind farm location

- Demonstrate an understanding of the basic components and function of a wind turbine

- Identify careers related to the wind energy industry

- Create a persuasive argument for a wind farm location

- Collaborate with peers to create a solution to a problem

TIME REQUIRED

- 8 days (approximately 45 minutes each day; see Tables 3.9–3.10, p. 43)

MATERIALS

Required Materials for Lesson 4

- STEM Research Notebooks

- Internet access for student research and viewing videos

- Handouts (attached at the end of this lesson)

- Materials students bring from home to create models (Make sure materials brought from home are safe to use in the activity.)
- Safety glasses or goggles

SAFETY NOTES

1. All laboratory occupants must wear safety glasses or googles during all phases of this inquiry activity.

2. Keep away from electrical sources when working with water because of the shock hazard.

3. Use caution when operating electrical devices (e.g., fans) because of the potential shock hazard, especially near water.

4. Use caution when working with sharps (scissors, sticks, stirrers, fan blades, etc.) to avoid cutting or puncturing skin.

5. Make sure all materials are put away after completing the activity.

6. Wash hands with soap and water after completing the activity.

CONTENT STANDARDS AND KEY VOCABULARY

Table 4.12 lists the content standards from the *NGSS, CCSS,* and the Framework for 21st Century Learning that this lesson addresses, and Table 4.13 (p. 175) presents the key vocabulary. Vocabulary terms are provided for both teacher and student use. Teachers may choose to introduce some or all of the terms to students.

Table 4.12. Content Standards Addressed in STEM Road Map Module Lesson 4

NEXT GENERATION SCIENCE STANDARDS

PERFORMANCE EXPECTATIONS

- 3-5-ETS1-2. Generate and compare multiple possible solutions to a problem based on how well each is likely to meet the criteria and constraints of the problem.

- 3-5-ETS1-3. Plan and carry out fair tests in which variables are controlled and failure points are considered to identify aspects of a model or prototype that can be improved.

DISCIPLINARY CORE IDEAS

ESS3.A. Natural Resources

- Energy and fuels that humans use are derived from natural sources, and their use affects the environment in multiple ways. Some resources are renewable over time, and others are not.

ETS1.A. Defining and Delimiting Engineering Problems

- Possible solutions to a problem are limited by available materials and resources (constraints). The success of a designed solution is determined by considering the desired features of a solution (criteria). Different proposals for solutions can be compared on the basis of how well each one meets the specified criteria for success or how well each takes the constraints into account. (3-5-ETS1-1)

CROSSCUTTING CONCEPTS

Systems and System Models

- A system is a group of related parts that make up a whole and can carry out functions its individual parts cannot.

- A system can be described in terms of its components and their interactions.

Energy and Matter

- Energy can be transferred in various ways and between objects.

Influence of Science, Engineering, and Technology on Society and the Natural World

- People's needs and wants change over time, as do their demands for new and improved technologies.

- Engineers improve existing technologies or develop new ones to increase their benefits, decrease known risks, and meet societal demands.

SCIENCE AND ENGINEERING PRACTICES

Developing and Using Models

- Identify limitations of models.

- Collaboratively develop and/or revise a model based on evidence that shows the relationships among variables for frequent and regular occurring events.

Table 4.12. (*continued*)

Developing and Using Models (continued)

- Develop and/or use models to describe and/or predict phenomena.

- Develop a diagram or simple physical prototype to convey a proposed object, tool, or process.

- Use a model to test cause and effect relationships or interactions concerning the functioning of a natural or designed system.

COMMON CORE STATE STANDARDS FOR MATHEMATICS

MATHEMATICAL PRACTICES

- MP1. Make sense of problems and persevere in solving them.

- MP2. Reason abstractly and quantitatively.

- MP3. Construct viable arguments and critique the reasoning of others.

- MP4. Model with mathematics.

- MP5. Use appropriate tools strategically.

MATHEMATICAL CONTENT

- 5.NBT.B.5. Fluently multiply multi-digit whole numbers using the standard algorithm.

- 5.MD.A.1. Convert among different-sized standard measurement units within a given measurement system (e.g., convert 5 cm to 0.05 m), and use these conversions in solving multi-step, real world problems.

COMMON CORE STATE STANDARDS FOR ENGLISH LANGUAGE ARTS

READING STANDARDS

- RI.5.1. Quote accurately from a text when explaining what the text says explicitly and when drawing inferences from the text.

- RI.5.4. Determine the meaning of general academic and domain-specific words and phrases in a text relevant to a grade 5 topic or subject area.

- RI.5.7. Draw on information from multiple print or digital sources, demonstrating the ability to locate an answer to a question quickly or to solve a problem efficiently.

- RI.5.9. Integrate information from several texts on the same topic in order to write or speak about the subject knowledgeably.

- RF.5.3. Know and apply grade-level phonics and word analysis skills in decoding words.

- RF.5.4. Read with sufficient accuracy and fluency to support comprehension.

- RF.5.4.a. Read grade-level text.

WRITING STANDARDS

- W.5.2. Write informative/explanatory texts to examine a topic and convey ideas and information clearly.

- W.5.4. Produce clear and coherent writing in which the development and organization are appropriate to task, purpose, and audience.

Table 4.12. (*continued*)

WRITING STANDARDS (*continued*)

- W.5.6. With some guidance and support from adults, use technology, including the Internet, to produce and publish writing as well as to interact and collaborate with others; demonstrate sufficient command of keyboarding skills to type a minimum of two pages in a single sitting.

- W.5.7. Conduct short research projects that use several sources to build knowledge through investigation of different aspects of a topic.

- W.5.8. Recall relevant information from experiences or gather relevant information from print and digital sources; summarize or paraphrase information in notes and finished work, and provide a list of sources.

SPEAKING AND LISTENING STANDARDS

- SL.5.1. Engage effectively in a range of collaborative discussions (one-on-one, in groups, and teacher-led) with diverse partners on grade 5 topics and texts, building on others' ideas and expressing their own clearly.

- SL.5.1.b. Follow agreed-upon rules for discussions and carry out assigned roles.

- SL.5.1d. Review the key ideas expressed and draw conclusions in light of information and knowledge gained from the discussions.

- SL.5.4. Report on a topic or text or present an opinion, sequencing ideas logically and using appropriate facts and relevant, descriptive details to support main ideas or themes; speak clearly at an understandable pace.

- SL.5.5. Include multimedia components (e.g., graphics, sound) and visual displays in presentations when appropriate to enhance the development of main ideas or themes.

- SL.5.6. Adapt speech to a variety of contexts and tasks, using formal English when appropriate to task and situation.

FRAMEWORK FOR 21ST CENTURY LEARNING

Interdisciplinary themes (global awareness; financial, economic, and business literacy; environmental literacy); Learning and Innovation Skills; Information, Media and Technology Skills; Life and Career Skills

---

Table 4.13. Key Vocabulary in Lesson 4

Key Vocabulary	Definition
economic development	efforts to improve the quality of life of an area by attracting businesses, creating jobs, and encouraging innovation and entrepreneurship
percentage	the amount of something in each hundred
persuasive language	the use of language to influence people to change their minds about something or agree with your opinion
proposal	a document that is used to persuade potential buyers, clients, or investors that a business offers a desirable product

TEACHER BACKGROUND INFORMATION

In this lesson, students first add to their understanding of wind and wind technology by investigating the implications of a wind farm for a community. In particular, students consider ways wind farms might benefit businesses and property owners locally. After students have gathered this information, they work in their teams to address the final challenge using information and artifacts from this and previous lessons to complete their wind farm plans and create persuasive presentations.

The Impacts of Wind Farms on Communities

Although wind farms have some environmental implications that may influence communities, as students explored in Lesson 3, they also can have positive impacts on communities. Potential benefits of wind farms include job creation, lower electricity costs, a reliable energy supply, stable energy prices, and cleaner air and water than with the use of fossil fuels for energy. Additionally, localities may benefit from taxes paid by wind farm owners and the reduced need to purchase fuels from distant places. More information about the benefits of wind farms to local communities can be found at *www.ucsusa. org/clean-energy/northeastern-states/community-benefits#.Vh1l88teKfQ*. While the focus of this lesson is on identifying community economic benefits of wind farms, students may identify economic disadvantages as well, although these are generally more difficult to substantiate than the benefits. Students should consider the full range of advantages and disadvantages, including economic and environmental factors, in their final presentations.

Careers in Wind Energy

For more information on careers in wind energy, see the following websites:

- *www.bls.gov/green/wind_energy*
- *https://energy.gov/eere/education/explore-careers-wind-power-0*

COMMON MISCONCEPTIONS

In this lesson, students synthesize their learning from previous lessons. See the tables on misconceptions in the earlier lessons if it becomes necessary to address any again during this final lesson.

PREPARATION FOR LESSON 4

Review the Teacher Background Information provided, assemble the materials for the lesson, and preview the videos recommended in the Learning Plan Components section below. Students will draw on their learning from previous lessons, relying on their

STEM Research Notebooks as a database for parts of the challenge. You should therefore ensure that students have placed all their work in their notebooks before beginning to work on the challenge.

Students will use simple materials that they bring from home to create a model of the device or plan they came up with to mitigate environmental impacts of their wind farm. Prompt your students on day 1 to consider what materials they will need, and have them create a plan for assembling the materials so that they will bring these to class for day 2 of this lesson.

LEARNING PLAN COMPONENTS
Introductory Activity/Engagement

Connection to the Challenge: In this lesson, students actively address the driving question for the module and challenge: Where could we locate a wind farm that a community would support? Begin each day by having students summarize their work from previous days' lessons and describe how their work contributed to answering the driving question and creating a solution for the module challenge.

Social Studies Class: Reintroduce the challenge by showing students the Wind Farm Challenge illustration (attached at the end of this lesson). Tell them that in this lesson, they will work in their teams to complete their challenge, bringing together their knowledge from previous lessons to create proposals for their wind farms. Tell students to imagine that they want to open a business, and ask what they think is the first thing they need to do. Students may offer ideas such as acquiring a building and making a product, but you should lead students to understand that making a plan is an essential step before starting a business. Tell students that they will need to convince community members and local government officials that their plan for a wind farm is good for the community, and they will need to make this argument based on their plan. Ask students to brainstorm about what they think should go into a plan for a wind farm. Create a class list of their ideas.

Science Class: Ask students to share what they learned about wind turbine blades from the Catch the Wind activity in Lesson 3. Were there any materials or blade lengths that worked better than others? Ask students to consider how blade design might affect the problem of birds being killed by wind turbines. Have students brainstorm about how a wind turbine could be designed to be safer for birds. Create a class list of ideas. Tell students that this is an example of how technology changes and improves in response to needs. Show the following three videos that illustrate various bird-safe wind turbines: "Bladeless Wind Generator Safe for Birds," at *https://learningenglish.voanews.com/a/blade-less-wind-power-generator-safe-for-birds/2856707.html*; "Bird Friendly Wind Turbines by Metric Motion Renewable Energy," at *https://www.youtube.com/watch?v=rWC6kjKI3GU*;

Wind Energy, Grade 5

and "EcoVert 75 Bird Friendly Vertical Axis Wind Turbine by Inerjy," at *www.youtube. com/watch?v=hoJZ0zdAePY*. As students watch, have them answer the following questions about each turbine design:

- What are this turbine's advantages?

- What are its disadvantages?

- What parts do you see on the turbine? (You may wish to have students make a sketch.)

Mathematics Connection: Show students the maps of U.S. wind power capacity by state for 2004 and 2014 (attached at the end of this lesson). Ask students if they think that wind capacity increased, decreased, or stayed the same over that 10-year period. Choose several states, and have students use subtraction to calculate the increase in wind capacity for those states. Tell students that the news often reports changes such as this as percentages. If students are not familiar with the concept of percentages, introduce this concept. For a video to support student understanding of percentages, see "Math Antics—What Are Percentages" (visit YouTube and search for this title or access the video directly at *www.youtube.com/watch?v=JeVSmq1Nrpw*). Have students calculate percentage changes in wind energy production during this time period for a variety of states and for the United States overall.

ELA Connection: Introduce the elements of a persuasive argument by showing the video "Intro to Persuasive Text Using TV Commercials," which uses television commercials to illustrate persuasive text (visit YouTube and search for this title or access the video directly at *www.youtube.com/watch?v=azttKmT0rVc*). Hold a class discussion to identify what elements of the commercials engaged viewers, and create a class list of the elements of persuasive text.

Activity/Exploration

This lesson's activity is the Wind Farm Challenge. Because this final challenge incorporates students' learning from all subject areas in the module, this section is presented as an integrated project, with the challenge parameters included in a combined social studies and science class. The tasks associated with the challenge can be divided among disciplinary areas if you prefer.

Social Studies and Science Class: Students have already compiled much of the necessary information for their presentations in previous lessons and will use their STEM Research Notebooks as a resource to access this information. The content focus in this lesson will be on investigating careers related to wind energy and the local economic implications of wind farms. Student teams will synthesize this new information with their existing knowledge to create their presentations. Because this is a complex activity, it is recommended that you have students break it down into daily tasks following the schedule given for Days 18–25 in Tables 3.9 and 3.10 (see p. 43). Additional details on each day's tasks are given below.

Day 18

Have student teams investigate the economic implications of wind farms for communities and explore careers related to wind energy using the Wind Farm Challenge Economic Benefits and Careers handouts attached at the end of this lesson. (You may wish to direct students to the websites given in the Teacher Background Information section on p. 176.) Each team should identify a town or city close to its proposed wind farm location using either road maps or a web mapping service such as Google Maps.

Day 19

Have students complete their research on careers. Then, using the materials that they brought from home, students should create a model of the device, product, or plan they proposed in Lesson 3 to mitigate the environmental impacts of wind farms. The models need not be functional (e.g., if they design a new blade, it does not have to be part of a functioning wind turbine), but rather will serve as models for marketing purposes for teams' proposals and will be used to address and alleviate concerns over environmental impacts of wind farms.

Day 20

Each student team should begin work on creating a plan or proposal for its wind farm that includes the following elements in this order:

1. Summary—Give a brief overview of the plan.

2. Description—Describe the location and what is being proposed.

3. Marketing—Explain the benefits of wind energy, why this location is desirable, and benefits to the local community.

4. Careers—Highlight one career in wind energy.

5. Setup—Address objections to wind energy and present models.

6. Financial plan—Present the budget.

7. Conclusion—Summarize the benefits of this plan.

The presentation should be targeted toward the local community near their wind farm site as well as potential investors. Have students use the Wind Farm Challenge Build a Proposal graphic organizers (attached at the end of this lesson), which include handouts for the various elements of the proposal listed above, to move through this task. You may wish to have team members work in pairs or separately on parts of the proposal, so that each team member becomes an "expert" on certain facets of the proposal. Using these handouts will help students organize their information and consider how they might present this information in a way that is compelling and persuasive to their audience of local community members and potential investors.

Day 21

Have teams finish organizing their information. Then, each team should decide on a presentation format and begin to create the presentation. It can be a video recording or an oral presentation using visual aids or PowerPoint.

Days 22–23

Student teams continue working on their presentations, which must be completed by the end of Day 23.

Days 24–25

Have student teams present their proposals to the class. You may wish to invite guests to act as fictional community members and investors. Allow the audience to ask questions after each presentation. Rubrics are attached at the end of this lesson.

Mathematics Connection: Budgeting (included in Wind Farm Challenge).

ELA Connection: Writing persuasive text and presentation skills (included in Wind Farm Challenge).

Explanation

Social Studies and Science Class: Students should understand that one of the basic principles underlying the field of economic development is that when businesses locate within a community, the community can experience a number of benefits. Among these are job growth, a higher tax base, and business provided to local companies during the construction and maintenance phases. With wind farms, there is an additional benefit of a locally produced, clean energy source with relatively few environmental impacts.

Review the EDP with students as they prepare to create their environmental impact mitigation models. Since students are creating models of products or systems that need

not function, the "test" phase of the EDP may be limited to getting feedback from peers or family members about their models.

Mathematics Connection: Not applicable.

ELA Connection: Not applicable.

Elaboration/Application of Knowledge

Social Studies and Science Class: Have students identify the potential for wind energy in their own communities. If there are wind farms nearby, arrange a field trip to a local wind farm. If there are no wind farms in your area, have students identify the reasons for the lack of wind farms. If your area has adequate winds, you might also ask them to propose a location for a wind farm.

STEM Research Notebook

Students should respond to the following prompt in their Research Notebooks: *Think about the alternative wind turbines you saw in the video you watched in the introduction to this lesson. Which of these turbines do you think is the best? Why? List as many benefits to this type of turbine as you can. Next, draw a diagram of the turbine you chose and label the parts. Can you think of any improvements to the design? Mark your improvements on your diagram.*

Mathematics Connection: Not applicable.

ELA Connection: Not applicable.

Evaluation/Assessment

Students may be assessed on the following performance tasks and other measures listed.

Performance Tasks

- Wind Farm Challenge Economic Benefits handout
- Wind Farm Challenge Careers handout
- Wind Farm Challenge presentation planning graphic organizers
- Wind Farm Challenge Presentation Rubric
- Wind Farm Challenge Model Design Rubric
- Wind Farm Challenge Individual Student Handout Rubric

Other Measures

- STEM Research Notebook Entry Rubric (see p. 88)
- Wind Farm Collaboration Rubric

INTERNET RESOURCES

Benefits of wind farms to local communities
- *www.ucsusa.org/clean-energy/northeastern-states/community-benefits#.Vh1l88teKfQ*

Information on careers in wind energy
- *www.bls.gov/green/wind_energy/*
- *https://energy.gov/eere/education/explore-careers-wind-power-0*

"Bladeless Wind Generator Safe for Birds" video
- *https://learningenglish.voanews.com/a/bladeless-wind-power-generator-safe-for-birds/2856707.html*

"Bird Friendly Wind Turbines by Metric Motion Renewable Energy" video
- *www.youtube.com/watch?v=rWC6kjKI3GU*

"EcoVert 75 Bird Friendly Vertical Axis Wind Turbine by Inerjy" video
- *www.youtube.com/watch?v=hoJZ0zdAePY*

"Math Antics—What Are Percentages" video
- *www.youtube.com/watch?v=JeVSmq1Nrpw*

"Intro to Persuasive Text Using TV Commercials" video
- *www.youtube.com/watch?v=azttKmT0rVc*

STUDENT HANDOUT

U.S. WIND POWER CAPACITY BY STATE, 2004

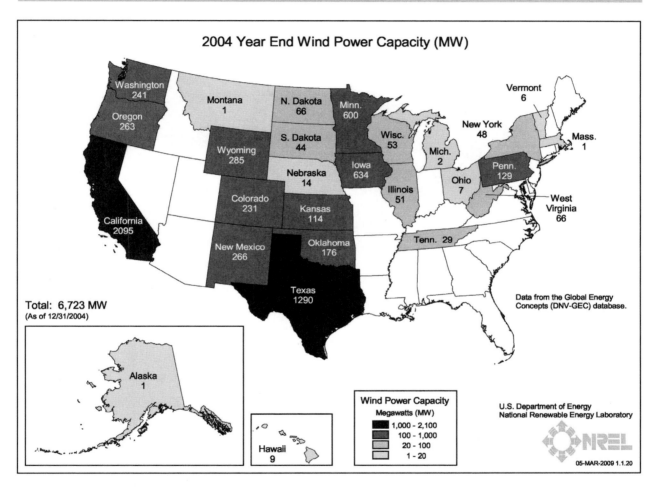

2004 Year End Wind Power Capacity (MW)

Total: 6,723 MW
(As of 12/31/2004)

Data from the Global Energy Concepts (DNV-GEC) database.

Wind Power Capacity
Megawatts (MW)
- 1,000 – 2,100
- 100 – 1,000
- 20 – 100
- 1 – 20

U.S. Department of Energy
National Renewable Energy Laboratory

NREL
05-MAR-2009 1.1.20

Source: National Renewable Energy Laboratories, U.S. Department of Energy. Public domain. *http://apps2.eere.energy.gov/wind/windexchange/wind_installed_capacity.asp.*

Note: A full-color version of this image is available on the book's Extras page at *www.nsta.org/roadmap-wind.*

STUDENT HANDOUT

U.S. WIND POWER CAPACITY BY STATE, 2014

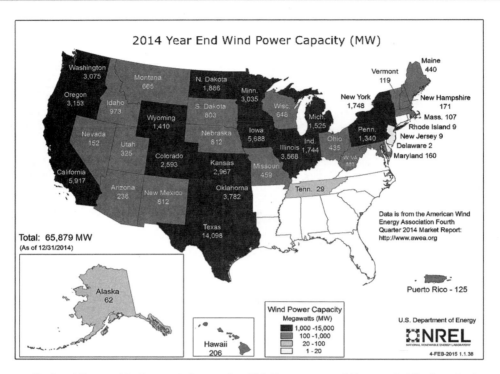

Source: National Renewable Energy Laboratories, U.S. Department of Energy. Public domain. *http:// apps2.eere.energy.gov/wind/windexchange/wind_installed_capacity.asp.*

Note: A full-color version of this image is available on the book's Extras page at *www.nsta.org/ roadmap-wind.*

STUDENT PACKET, PAGE 1

WIND FARM CHALLENGE

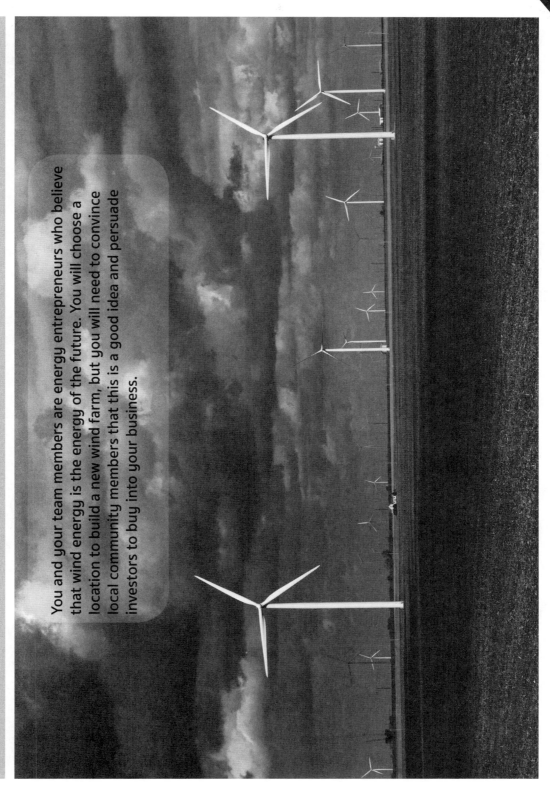

You and your team members are energy entrepreneurs who believe that wind energy is the energy of the future. You will choose a location to build a new wind farm, but you will need to convince local community members that this is a good idea and persuade investors to buy into your business.

Name: _____   Wind Farm Name: _____

ECONOMIC BENEFITS: WIND FARM CHALLENGE

As entrepreneurs proposing a wind farm in a community, you must highlight the benefits of your farm for the community in which you plan to build it.

Using a road map or a web mapping service such as Google Maps, find the town that is

closest to your proposed wind farm location: _____

Use the internet to research the benefits of wind farms to a community. Choose three to include in your proposal and describe them here:

Benefit 1:

This will be good for this community because:

Benefit 2:

This will be good for this community because:

Benefit 3:

This will be good for this community because:

4

Name: _____ Wind Farm Name: _____

CAREERS: WIND FARM CHALLENGE

When a new business such as a wind farm moves to a community, people are interested in what sorts of jobs it might create locally. Your task is to identify three jobs related to wind farms and describe them here.

Job 1:

Describe the work this person does:

What kind of education or training does someone need to do this job?

Job 2:

Describe the work this person does:

What kind of education or training does someone need to do this job?

Job 3:

Describe the work this person does:

What kind of education or training does someone need to do this job?

4

Name: _____ Wind Farm Name: _____

PRESENTATION GRAPHIC ORGANIZER: SUMMARY AND DESCRIPTION

WIND FARM CHALLENGE

Introduce Your Wind Farm Team

Team information to include:

Presentation features to include (e.g., music or images):

Introduce Your Location

Wind farm location (be sure to show the location on a map):

Acreage and number of turbines planned:

Overview of what your team plans to do:

Presentation features to include (e.g., music or images):

Name: _____   Wind Farm Name: _____

STUDENT PACKET, PAGE 5

PRESENTATION GRAPHIC ORGANIZER: MARKETING WIND FARM CHALLENGE

Benefits of Wind Energy and Wind Farms

Presentation features to include (e.g., music or images):

Benefits of wind energy overall:

Benefits of a wind farm to this community:

Benefits of the Proposed Location

Presentation features to include (e.g., music or images):

Why is this a good spot for a wind farm?

What will be the impacts of choosing this location on the community?

What other benefits can you think of that would convince this community that the wind farm is a good idea?

Name: _____   Wind Farm Name: _____

PRESENTATION GRAPHIC ORGANIZER: SETUP WIND FARM CHALLENGE

Objections to Wind Energy and Wind Farms

Presentation features to include (e.g., music or images):

What are the objections to wind energy and wind farms?

Your Solution

Presentation features to include (how will you present your model?):

What is your solution to address the environmental impacts?

How will this make wind energy safer, cleaner, or better?

Is there anything else the community and your potential investors should know about how you plan to set up the wind farm?

Name: _____ Wind Farm Name: _____

PRESENTATION: FINANCIAL PLAN
WIND FARM CHALLENGE

Financial Plan

How will you present your budget?

Why should investors feel confident about your budget?

Presentation features to include (e.g., music or images):

Looking Ahead

How will your budget change in the future?

What income will you have?

Name: _____

Wind Farm Challenge Individual Student Handouts Rubric (20 points possible)

Handout	Below Standard (0–1)	Approaching Standard (2–3)	Meets or Exceeds Standard (4–5)	Student Score
ECONOMIC BENEFITS	• Only 1 or 2 benefits are listed, or the explanations of how these things will benefit the community are missing or incomplete.	• All 3 benefits are listed; however, the explanation of how these things will benefit the community may be difficult to understand, contain grammatical errors, or be incomplete.	• All 3 benefits are listed and have clear, understandable explanations that contain few or no grammatical errors.	
CAREERS	• Only 1 or 2 careers are listed, or the education and training information is missing or incomplete.	• All 3 careers are listed; however, the education and training information may contain grammatical errors or be incomplete.	• All 3 careers are listed, with complete education and training information that contains few or no grammatical errors.	
PRESENTATION PLAN (SUMMARY, MARKETING)	• Significant components of the plan are missing, and there is little evidence that student understands the benefits and other implications of wind energy.	• Most components of the plan are present; however, some information is missing or incorrect.	• All components of the plan are present, and there is evidence that student understands the benefits and other implications of wind energy.	
PRESENTATION PLAN (SETUP AND FINANCIAL PLAN)	• Significant components of plan are missing, and there is little evidence that student understands the environmental impacts and other costs of wind energy.	• Most components of the plan are present; however, some information is missing or incorrect.	• All components of the plan are present, and there is evidence that student understands the environmental impacts and other costs of wind energy.	

TOTAL SCORE: _____

Team Name: _____

Wind Farm Challenge Presentation Rubric (30 points possible)

Team Performance	Below Standard (0–2)	Approaching Standard (3–4)	Meets or Exceeds Standard (5–6)	Team Score
INFORMATION	• Team includes little interesting or factual information and may omit important information.	• Team includes some interesting and factual information.	• Team includes complete, interesting, and factual information.	
IDEAS AND ORGANIZATION	• Team does not have a main idea or organizational strategy. • Presentation does not include an introduction or conclusion. • Presentation is confusing and uninformative. • Team uses presentation time poorly and it is too short or too long.	• Team has a main idea and organizational strategy, although it may not be clear. • Presentation includes either an introduction or conclusion or both. • Presentation is fairly coherent, well organized, and informative. • Team uses presentation time adequately, but presentation may be slightly too short or too long.	• Team has a clear main idea and organizational strategy. • Presentation includes both an introduction and conclusion. • Presentation is coherent, well organized, and informative. • Team uses presentation time well, and presentation is an appropriate length.	

Wind Farm Challenge Presentation Rubric (*continued*)

Team Performance	Below Standard (0–2)	Approaching Standard (3–4)	Meets or Exceeds Standard (5–6)	Team Score
PRESENTATION STYLE	• Only one or two team members participate in the presentation. • Presenters are difficult to understand. • Presenters use language inappropriate for audience (slang, poor grammar, frequent filler words such as "uh," "um").	• Some, but not all, team members participate in the presentation. • Most presenters are understandable, but volume may be too low or some presenters may mumble. • Presenters use some language inappropriate for audience (slang or poor grammar, some use of filler words such as "uh," "um").	• All team members participate in the presentation. • Presenters are easy to understand. • Presenters use appropriate language for audience (no slang or poor grammar, infrequent use of filler words such as "uh," "um").	
VISUAL AIDS	• Team does not use any visual aids to presentation • Visual aids are used but do not add to the presentation	• Team uses some visual aids to presentation, but they may be poorly executed or distract from the presentation.	• Team uses well-produced visual aids or media that clarify and enhance presentation.	
RESPONSE TO AUDIENCE QUESTIONS	• Team fails to respond to questions from audience or responds inappropriately.	• Team responds appropriately to audience questions, but responses may be brief, incomplete, or unclear.	• Team responds clearly and in detail to audience questions and seeks clarification of questions.	

TOTAL SCORE: _____

Student Name _____ Team Name _____

Wind Farm Challenge Collaboration Rubric (30 points possible)

Individual Performance	Below Standard (0–3)	Approaching Standard (4–7)	Meets or Exceeds Standard (8–10)	Student Score
INDIVIDUAL ACCOUNTABILITY	• Student is unprepared. • Student does not communicate with team members and does not manage tasks as agreed on by the team. • Student does not complete or participate in project tasks. • Student does not complete tasks on time. • Student does not use feedback from others to improve work.	• Student is usually prepared. • Student sometimes communicates with team members and manages tasks as agreed on by the team, but not consistently. • Student completes or participates in some project tasks but needs to be reminded. • Student completes most tasks on time. • Student sometimes uses feedback from others to improve work.	• Student is consistently prepared. • Student consistently communicates with team members and manage tasks as agreed on by the team. Student discusses and reflects on ideas with the team. • Student completes or participates in project tasks without being reminded. • Student completes tasks on time. • Student uses feedback from others to improve work.	
TEAM PARTICIPATION	• Student does not help the team solve problems; may interfere with teamwork. • Student does not express ideas clearly, pose relevant questions, or participate in group discussions. • Student does not give useful feedback to other team members. • Student does not volunteer to help others when needed.	• Student cooperates with the team but may not actively help solve problems. • Student sometimes expresses ideas, poses relevant questions, elaborates in response to questions, and participates in group discussions. • Student provides some feedback to team members. • Student sometimes volunteers to help others.	• Student helps the team solve problems and manage conflicts. • Student makes discussions effective by clearly expressing ideas, posing questions, and responding thoughtfully to team members' questions and perspectives. • Student gives useful feedback to others so they can improve their work. • Student volunteers to help others if needed.	

Wind Farm Challenge Collaboration Rubric (*continued*)

Individual Performance	Below Standard (0–3)	Approaching Standard (4–7)	Meets or Exceeds Standard (8–10)	Student Score
PROFESSIONALISM AND RESPECT FOR TEAM MEMBERS	• Student is impolite or disrespectful to other team members. • Student does not acknowledge or respect others' ideas and perspectives.	• Student is usually polite and respectful to other team members. • Student usually acknowledges and respects others' ideas and perspectives.	• Student is consistently polite and respectful to other team members. • Student consistently acknowledges and respects others' ideas and perspectives.	

TOTAL SCORE: _____

Team Name: _____

Wind Farm Challenge Model Design Rubric (10 points possible)

Team Performance	Below Standard (0-1)	Approaching Standard (2-3)	Meets or Exceeds Standard (4-5)	Team Score
CREATIVITY AND INNOVATION	• Design reflects little creativity with use of materials, shows lack of understanding of project purpose, and has no innovative design features. • Design is impractical. • Design has several elements that do not fit.	• Design reflects some creativity with use of materials, shows a basic understanding of project purpose, and has limited innovative design features. • Design is limited in practicality. • Design has some interesting elements but may be excessive or inappropriate.	• Design reflects creative use of materials, shows a sound understanding of project purpose, and has distinct innovative design features. • Design is practical. • Design is well crafted and includes interesting elements that are appropriate for the purpose.	
CONCEPTUAL UNDERSTANDING	• Design incorporates no or few features that reflect understanding of concepts.	• Design incorporates some features that reflect a limited understanding of concepts.	• Design incorporates several features that reflect a sound understanding of concepts.	

TOTAL SCORE: _____

Wind Energy, Grade 5

STUDENT HANDOUT

TOPOGRAPHIC MAP, SALT LAKE CITY, UTAH

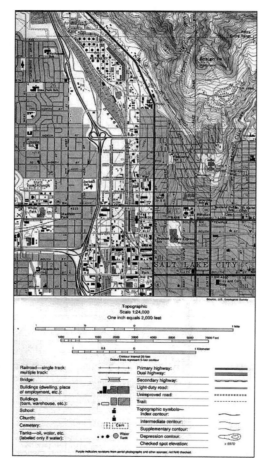

Source: U.S. Geological Survey. Public domain. *http://egsc.usgs.gov/isb//pubs/teachers-packets/mapshow/topo.html.*

Note: A full-color version of this image is available on the book's Extras page at *www.nsta.org/roadmap-wind.*

4

STUDENT HANDOUT

SHADED RELIEF MAP OF SALT LAKE CITY, UTAH

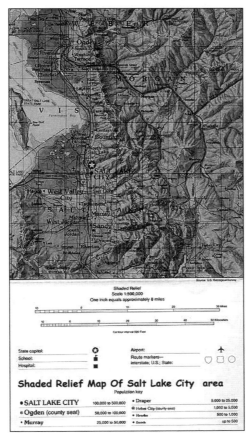

Source: U.S. Geological Survey. Public domain. *http://egsc.usgs.gov/isb//pubs/teachers-packets/ mapshow/relief.html.*

Note: A full-color version of this image is available on the book's Extras page at *www.nsta.org/ roadmap-wind.*

TRANSFORMING LEARNING WITH WIND ENERGY AND THE *STEM ROAD MAP CURRICULUM SERIES*

Carla C. Johnson

This chapter serves as a conclusion to the Wind Energy integrated STEM curriculum module, but it is just the beginning of the transformation of your classroom that is possible through use of the *STEM Road Map Curriculum Series*. In this book, many key resources have been provided to make learning meaningful for your students through integration of science, technology, engineering, and mathematics, as well as social studies and English language arts, into powerful problem- and project-based instruction. First, the Wind Energy curriculum is grounded in the latest theory of learning for children in elementary school specifically. Second, as your students work through this module, they engage in using the engineering design process (EDP) and build prototypes like engineers and STEM professionals in the real world. Third, students acquire important knowledge and skills grounded in national academic standards in mathematics, English language arts, science, and 21st century skills that will enable their learning to be deeper, retained longer, and applied throughout, illustrating the critical connections within and across disciplines. Finally, authentic formative assessments, including strategies for differentiation and addressing misconceptions, are embedded within the curriculum activities.

The Wind Energy curriculum in the Innovation and Progress STEM Road Map theme can be used in single-content elementary school classrooms (e.g., science) where there is only one teacher or expanded to include multiple teachers and content areas across classrooms. Through the exploration of the Wind Farm Challenge, students engage in a real-world STEM problem on the first day of instruction and gather necessary knowledge and skills along the way in the context of solving the problem.

The other topics in the *STEM Road Map Curriculum Series* are designed in a similar manner, and NSTA Press plans to publish additional volumes in this series for this and other grade levels. The tentative list of books includes the following themes and subjects:

- Innovation and Progress
  - Amusement park of the future
  - Construction materials
  - Harnessing solar energy
  - Transportation in the future
- The Represented World
  - Rainwater analysis
  - Recreational STEM: Swing set makeover
- Sustainable Systems
  - Composting: reduce, reuse, recycle
  - Hydropower efficiency
- Optimizing the Human Condition
  - Water conservation: Think global, act local

If you are interested in professional development opportunities focused on the STEM Road Map specifically or integrated STEM or STEM programs and schools overall, contact the lead editor of this project, Dr. Carla C. Johnson (*carlacjohnson@purdue.edu),* associate dean and professor of science education at Purdue University. Someone from the team will be in touch to design a program that will meet your individual, school, or district needs.

APPENDIX

CONTENT STANDARDS ADDRESSED IN THIS MODULE

NEXT GENERATION SCIENCE STANDARDS

Table A1 (p. 204) lists the science and engineering practices, disciplinary core ideas, and crosscutting concepts this module adresses. The supported performance expectations are as follows:

- 5-ESS2-1. Develop a model using an example to describe ways the geosphere, biosphere, hydrosphere, and/or atmosphere interact.

- 3-5-ETS1-1. Define a simple design problem reflecting a need or a want that includes specified criteria for success and constraints on materials, time, or cost.

- 3-5-ETS1-2. Generate and compare multiple possible solutions to a problem based on how well each is likely to meet the criteria and constraints of the problem.

- 3-5-ETS1-3. Plan and carry out fair tests in which variables are controlled and failure points are considered to identify aspects of a model or prototype that can be improved.

Table A1. *Next Generation Science Standards (NGSS)*

Science and Engineering Practices

ASKING QUESTIONS AND DEFINING PROBLEMS

- Ask questions about what would happen if a variable is changed.
- Identify scientific (testable) and non-scientific (non-testable) questions.
- Ask questions that can be investigated and predict reasonable outcomes based on patterns such as cause and effect relationships.
- Use prior knowledge to describe problems that can be solved.
- Define a simple design problem that can be solved through the development of an object, tool, process, or system and includes several criteria for success and constraints on materials, time, or cost.

DEVELOPING AND USING MODELS

- Identify limitations of models.
- Collaboratively develop and/or revise a model based on evidence that shows the relationships among variables for frequent and regular occurring events.
- Develop a model using an analogy, example, or abstract representation to describe a scientific principle or design solution.
- Develop and/or use models to describe and/or predict phenomena.
- Develop a diagram or simple physical prototype to convey a proposed object, tool, or process.
- Use a model to test cause and effect relationships or interactions concerning the functioning of a natural or designed system.

PLANNING AND CARRYING OUT INVESTIGATIONS

- Plan and conduct an investigation collaboratively to produce data to serve as the basis for evidence, using fair tests in which variables are controlled and the number of trials considered.
- Evaluate appropriate methods and/or tools for collecting data.
- Make observations and/or measurements to produce data to serve as the basis for evidence for an explanation of a phenomenon or test a design solution.
- Make predictions about what would happen if a variable changes.
- Test two different models of the same proposed object, tool, or process to determine which better meets criteria for success.

ANALYZING AND INTERPRETING DATA

- Represent data in tables and/or various graphical displays (bar graphs, pictographs and/or pie charts) to reveal patterns that indicate relationships.
- Analyze and interpret data to make sense of phenomena, using logical reasoning, mathematics, and/or computation.

Table A1. (*continued*)

Science and Engineering Practices (*continued*)

ANALYZING AND INTERPRETING DATA (*continued*)

- Compare and contrast data collected by different groups in order to discuss similarities and differences in their findings.

- Analyze data to refine a problem statement or the design of a proposed object, tool, or process.

- Use data to evaluate and refine design solutions.

USING MATHEMATICAL AND COMPUTATIONAL THINKING

- Decide if qualitative or quantitative data are best to determine whether a proposed object or tool meets criteria for success.

- Organize simple data sets to reveal patterns that suggest relationships.

- Describe, measure, estimate, and/or graph quantities (e.g., area, volume, weight, time) to address scientific and engineering questions and problems.

- Create and/or use graphs and/or charts generated from simple algorithms to compare alternative solutions to an engineering problem.

CONSTRUCTING EXPLANATIONS AND DESIGNING SOLUTIONS

- Construct an explanation of observed relationships (e.g., the distribution of plants in the back yard).

- Use evidence (e.g., measurements, observations, patterns) to construct or support an explanation or design a solution to a problem.

- Identify the evidence that supports particular points in an explanation.

- Apply scientific ideas to solve design problems.

- Generate and compare multiple solutions to a problem based on how well they meet the criteria and constraints of the design solution.

ENGAGING IN ARGUMENT FROM EVIDENCE

- Compare and refine arguments based on an evaluation of the evidence presented.

- Distinguish among facts, reasoned judgment based on research findings, and speculation in an explanation.

- Respectfully provide and receive critiques from peers about a proposed procedure, explanation, or model by citing relevant evidence and posing specific questions.

- Construct and/or support an argument with evidence, data, and/or a model.

- Use data to evaluate claims about cause and effect.

- Make a claim about the merit of a solution to a problem by citing relevant evidence about how it meets the criteria and constraints of the problem.

Table A1. (*continued*)

Science and Engineering Practices (*continued*)

OBTAINING, EVALUATING, AND COMMUNICATING INFORMATION
- Read and comprehend grade-appropriate complex texts and/or other reliable media to summarize and obtain scientific and technical ideas and describe how they are supported by evidence.
- Compare and/or combine across complex texts and/or other reliable media to support the engagement in other scientific and/or engineering practices.
- Combine information in written text with that contained in corresponding tables, diagrams, and/or charts to support the engagement in other scientific and/or engineering practices.
- Obtain and combine information from books and/or other reliable media to explain phenomena or solutions to a design problem.
- Communicate scientific and/or technical information orally and/or in written formats, including various forms of media as well as tables, diagrams, and charts.

Disciplinary Core Ideas

ESS3.A. NATURAL RESOURCES
- Energy and fuels that humans use are derived from natural sources, and their use affects the environment in multiple ways. Some resources are renewable over time, and others are not.

ETS1.A. DEFINING AND DELIMITING ENGINEERING PROBLEMS
- Possible solutions to a problem are limited by available materials and resources (constraints). The success of a designed solution is determined by considering the desired features of a solution (criteria). Different proposals for solutions can be compared on the basis of how well each one meets the specified criteria for success or how well each takes the constraints into account. (3-5-ETS1-1)
- Research on a problem should be carried out before beginning to design a solution. Testing a solution involves investigating how well it performs under a range of likely conditions. (3-5-ETS1-2)
- At whatever stage, communicating with peers about proposed solutions is an important part of the design process, and shared ideas can lead to improved designs. (3-5-ETS1-2)
- Tests are often designed to identify failure points or difficulties, which suggest the elements of the design that need to be improved (3-5-ETS1-3)

ETS1.C. OPTIMIZING THE DESIGN SOLUTION
- Different solutions need to be tested in order to determine which of them best solves the problem, given the criteria and the constraints. (3-5-ETS1-3)

Table A1. (*continued*)

Crosscutting Concepts

SCALE, PROPORTION, AND QUANTITY

- Standard units are used to measure and describe physical quantities such as weight, time, temperature, and volume.

SYSTEMS AND SYSTEM MODELS

- A system is a group of related parts that make up a whole and can carry out functions its individual parts cannot.

- A system can be described in terms of its components and their interactions.

ENERGY AND MATTER

- Energy can be transferred in various ways and between objects.

CAUSE AND EFFECT

- Cause and effect relationships are routinely identified, tested, and used to explain change.

STABILITY AND CHANGE

- Change is measured in terms of differences over time and may occur at different rates.

INFLUENCE OF SCIENCE, ENGINEERING, AND TECHNOLOGY ON SOCIETY AND THE NATURAL WORLD

- People's needs and wants change over time, as do their demands for new and improved technologies.

- Engineers improve existing technologies or develop new ones to increase their benefits, decrease known risks, and meet societal demands.

Table A2. Common Core Mathematics and English Language Arts (ELA) Standards

MATHEMATICAL PRACTICES	READING STANDARDS
• MP1. Make sense of problems and persevere in solving them.	• RI.5.1. Quote accurately from a text when explaining what the text says explicitly and when drawing inferences from the text.
• MP2. Reason abstractly and quantitatively.	• RI.5.4. Determine the meaning of general academic and domain-specific words and phrases in a text relevant to a grade 5 topic or subject area.
• MP3. Construct viable arguments and critique the reasoning of others.	• RI.5.7. Draw on information from multiple print or digital sources, demonstrating the ability to locate an answer to a question quickly or to solve a problem efficiently.
• MP4. Model with mathematics.	
• MP5. Use appropriate tools strategically.	• RI.5.9. Integrate information from several texts on the same topic in order to write or speak about the subject knowledgeably.
MATHEMATICAL CONTENT	• RF.5.3. Know and apply grade-level phonics and word analysis skills in decoding words.
• 5.NBT.B.5. Fluently multiply multi-digit whole numbers using the standard algorithm.	• RF.5.4. Read with sufficient accuracy and fluency to support comprehension.
• 5.MD.A.1. Convert among different-sized standard measurement units within a given measurement system (e.g., convert 5 cm to 0.05 m), and use these conversions in solving multi-step, real world problems.	• RF.5.4.a. Read grade-level text.
	WRITING STANDARDS
	• W.5.1. Write opinion pieces on topics or texts, supporting a point of view with reasons and information.
• 5.G.A.2. Represent real world and mathematical problems by graphing points in the first quadrant of the coordinate plane, and interpret coordinate values of points in the context of the situation.	• W.5.2. Write informative/explanatory texts to examine a topic and convey ideas and information clearly.
	• W.5.4. Produce clear and coherent writing in which the development and organization are appropriate to task, purpose, and audience.
	• W.5.6. With some guidance and support from adults, use technology, including the Internet, to produce and publish writing as well as to interact and collaborate with others; demonstrate sufficient command of keyboarding skills to type a minimum of two pages in a single sitting.

Table A2. (*continued*)

	WRITING STANDARDS (*continued*) • W.5.7. Conduct short research projects that use several sources to build knowledge through investigation of different aspects of a topic. • W.5.8. Recall relevant information from experiences or gather relevant information from print and digital sources; summarize or paraphrase information in notes and finished work, and provide a list of sources. **SPEAKING AND LISTENING STANDARDS** • SL.5.1. Engage effectively in a range of collaborative discussions (one-on-one, in groups, and teacher-led) with diverse partners on grade 5 topics and texts, building on others' ideas and expressing their own clearly. • SL.5.1.b. Follow agreed-upon rules for discussions and carry out assigned roles. • SL.5.1d. Review the key ideas expressed and draw conclusions in light of information and knowledge gained from the discussions. • SL.5.4. Report on a topic or text or present an opinion, sequencing ideas logically and using appropriate facts and relevant, descriptive details to support main ideas or themes; speak clearly at an understandable pace. • SL.5.5. Include multimedia components (e.g., graphics, sound) and visual displays in presentations when appropriate to enhance the development of main ideas or themes. • SL.5.6. Adapt speech to a variety of contexts and tasks, using formal English when appropriate to task and situation.

Table A3. 21st Century Skills From the Framework for 21st Century Learning

21st Century Skills	Learning Skills and Technology Tools	Teaching Strategies	Evidence of Success
INTERDISCIPLINARY THEMES • Global Awareness • Financial, Economic, and Business Literacy • Environmental Literacy	• Using 21st century skills to understand and address global issues. • Knowing how to make appropriate economic choices. • Understanding the role of the economy in society. • Using entrepreneurial skills to enhance work-place productivity and career options. • Demonstrate knowledge and understanding of the environment and the circumstances and conditions affecting it. • Demonstrate knowledge and understanding of society's impact on the natural world. • Investigate and analyze environmental issues, and make accurate conclusions about effective solutions.	• Lessons include investigations of renewable energy resources. • Have students consider the role of financial cost in their decision-making process. • Have students consider the environmental impacts of energy sources and the ways wind turbines can impact the environment.	• Students understand renewable energy resources, financial costs, and environmental impacts.
LEARNING AND INNOVATION SKILLS • Creativity and Innovation • Critical Thinking and Problem Solving • Communication and Collaboration	• Use a wide range of idea creation techniques (such as brainstorming) to create new ideas and refine and evaluate those ideas. • Be open and responsive to new and diverse perspectives; incorporate group input and feedback into the work. • Demonstrate originality and inventiveness in work and understand the real world limits to adopting new ideas.	• Use class and group brainstorming sessions and the EDP to provide students the opportunity to think originally and creatively while solving problems. • Students work as teams for a number of activities and for the final challenge in this module, and the class works together to create guidelines for collaboration.	• Students understand and implement the EDP. • Students work collaboratively and effectively in groups to solve a problem and complete a group project.

Table A3. (*continued*)

21st Century Skills	Learning Skills and Technology Tools	Teaching Strategies	Evidence of Success
LEARNING AND INNOVATION SKILLS (*continued*)	• Analyze how parts of a whole interact with each other to produce overall outcomes in complex systems. • Effectively analyze and evaluate evidence, arguments, claims and beliefs and draw conclusions. • Solve various problems in both conventional and innovative ways. • Articulate thoughts and ideas effectively using oral, written and nonverbal communication skills in a variety of forms and contexts. • Listen effectively to decipher meaning, including knowledge, values, attitudes and intentions. • Use communication for a range of purposes (e.g., to inform, instruct, motivate and persuade). • Demonstrate ability to work effectively and respectfully with diverse teams. • Exercise flexibility and willingness to be helpful in making necessary compromises to accomplish a common goal. • Assume shared responsibility for collaborative work, and value the individual contributions made by each team member.		

Table A3. (*continued*)

21st Century Skills	Learning Skills and Technology Tools	Teaching Strategies	Evidence of Success
INFORMATION, MEDIA AND TECHNOLOGY SKILLS • Information Literacy • Media Literacy • ICT Literacy	• Access information efficiently (time) and effectively (sources). • Evaluate information critically and competently. • Use information accurately and creatively for the issue or problem at hand. • Understand both how and why media messages are constructed, and for what purposes. • Use technology as a tool to research, organize, evaluate and communicate information.	• Students conduct internet research on various topics throughout the module. • Hold a class discussion of internet search strategies and website evaluation. • Have students use technology tools to create presentations and convey information.	• Students complete internet research in various activities. • Students give informative presentations to the class using technology resources.
LIFE AND CAREER SKILLS • Flexibility and Adaptability • Initiative and Self-Direction • Social and Cross Cultural Skills • Productivity and Accountability • Leadership and Responsibility	• Adapt to varied roles, jobs responsibilities, schedules and contexts. • Incorporate feedback effectively. • Deal positively with praise, setbacks and criticism. • Understand, negotiate and balance diverse views and beliefs to reach workable solutions, particularly in multi-cultural environments. • Balance tactical (short-term) and strategic (long-term) goals. • Utilize time and manage workload efficiently.	• Students' group work gives them opportunities to play various roles within groups. • The class collaboration guidelines list serves as a basis for guiding effective group work, including flexibility, initiative, and appropriate social behaviors. • Use the EDP to scaffold students' ability to manage time and prioritize tasks, test their ideas or solutions, and redesign those based on performance or feedback from others.	• Team projects are completed on time with evidence of participation by all team members. • Presentations use appropriate language and vocabulary and demonstrate a concerted and organized team effort.

Table A3. (*continued*)

21st Century Skills	Learning Skills and Technology Tools	Teaching Strategies	Evidence of Success
LIFE AND CAREER SKILLS (*continued*)	• Monitor, define, prioritize, and complete tasks without direct oversight. • Reflect critically on past experiences in order to inform future progress. • Know when it is appropriate to listen and when to speak. • Conduct themselves in a respectable, professional manner. • Use interpersonal and problem-solving skills to influence and guide others toward a goal. • Leverage strengths of others to accomplish a common goal.		

Table A4. English Language Development Standards

ELD STANDARD 1: SOCIAL AND INSTRUCTIONAL LANGUAGE

English language learners communicate for Social and Instructional purposes within the school setting.

ELD STANDARD 2: THE LANGUAGE OF LANGUAGE ARTS

English language learners communicate information, ideas, and concepts necessary for academic success in the content area of Language Arts.

ELD STANDARD 3: THE LANGUAGE OF MATHEMATICS

English language learners communicate information, ideas, and concepts necessary for academic success in the content area of Mathematics.

ELD STANDARD 4: THE LANGUAGE OF SCIENCE

English language learners communicate information, ideas, and concepts necessary for academic success in the content area of Science.

ELD STANDARD 5: THE LANGUAGE OF SOCIAL STUDIES

English language learners communicate information, ideas, and concepts necessary for academic success in the content area of Social Studies.

Source: WIDA, 2012, 2012 Amplification of the English language development standards: Kindergarten–grade 12, *www.wida.us/standards/eld.aspx.*

INDEX

Page numbers printed in **boldface type** indicate tables, figures, or handouts.

NATIONAL SCIENCE TEACHERS ASSOCIATION